20

Calculus

Calculus

For Business and Social Science

Pace University **William J. Adams**

Xerox College Publishing Lexington, Massachusetts / Toronto

To JOSEPH M. ADAMS and the memory of ANNA A. ADAMS

Preface

My objective in this book is to present a lucid exposition of topics in calculus for students who plan to pursue studies in business, social science, and other disciplines which may, directly or indirectly, bring them into contact with the calculus. Students of business and economics, for example, will find the contact direct; many of the basic concepts of these disciplines (marginal cost, marginal revenue, elasticity of demand, and so on) are defined in terms of the calculus. Students of education, political science, psychology, and sociology, on the other hand, may find the contact more remote; basic concepts in these disciplines are not defined in terms of the calculus. But the calculus is a prerequisite for an in-depth study of such disciplines as probability and statistics, which in turn have provided the social sciences with important methods. Moreover, a knowledge of basic calculus is a minimal prerequisite for contact with the now-small-but-growing disciplines of mathematical social science and mathematical psychology. My other concern is with the student who, apart from future course work and professional needs, wants to be a well-educated person. The calculus is one of the great triumphs of the human intellect and, like fine art and fine music, brings us together as human beings interested in partaking of the best of the human spirit.

To show that there are natural and important applications of calculus to business and social science, many of the applications presented in this book pertain to these two areas. At the same time it is not my view that applications to business are only for business students, that applications to physics are only for physics students, and so on. Regardless of professional interests, if a student experiences a wide spectrum of applications of the calculus, then the student will gain a deeper understanding of calculus and thus a broader appreciation of its power, scope, and limitations for his or her own professional area of study.

The book is written at a level suitable for beginning college students with minimal training in mathematics: the only background assumed is high school intermediate algebra or its equivalent. The student who feels uncertain about algebra—exponents, logarithms, and the like—may want to consult the algebra review presented in Appendix 3 when the situation seems appropriate. The book is intended for one-semester or two-quarter courses in calculus, or as part of a multi-semester sequence in mathematics of which calculus is a part. The book's organization permits a number of omissions and a choice of the order in which topics can be covered. The options available are indicated by the chart which follows the table of contents.

Of the postulates which guided the development of this book, the following may be of special interest.

1. *The book should be readable by students.* A special effort has been made to present the material in an intuitive and informal manner without sacrificing correctness. Concepts are introduced through examples, and rigorous proofs that are above the expected level of mathematical maturity of the intended audience have been avoided. Every effort has been made to avoid ambushing the reader with unforseen algebraic traps. As is well known, algebraic difficulties in illustrative examples and exercises can make the concept under discussion more complicated than it really is.

2. *Emphasis should be placed on realistic applications that are self-contained.* The intent of this text is to convey a sense of realism to the reader, to avoid the frequent "Alice in Wonderland" flavor that comes from an overcommitment to artificial applications.

3. *The mathematical machinery developed should be appropriate to the applications presented.* This text brings a sense of proportion to the mathematics developed and the applications presented. Every attempt has been made to leave the reader with the feeling that any extra effort required to master theory brings with it appropriate rewards in understanding and realistic applications.

4. *Answers to problems are not enough.* Answers to approximately half of the exercises in the text appear at the end of this book. But answers alone are not enough to indicate to the student where he has made mistakes or to show that a correct solution was reached through incorrect reasoning. Thus a companion Study Guide to the text has been prepared to include answers with detailed solutions to all problems in the book; it should be of benefit to both students and instructors.

Some comments on the scope, spirit, and technical features of the chapters may serve as a useful introduction to the text as a whole. Chapter 1 introduces the fundamental concepts of function and limit. It has been my experience that many students confuse $\lim_{x \to a} f(x)$ with $f(a)$: to such students every function is continuous. Thus in this book the limit concept is given a thorough, but intuitive, discussion.

In preparation for the formal introduction of the derivative in Chapter 2, limit problems with the derivative structure are introduced in a gentle way in the exercises of Chapter 1. This marks stage one in the introduction of the derivative.

The number e is introduced in Section 7 of Chapter 1 via a discussion of continuous compounding of interest. Such an introduction in a concrete setting considerably eases the sting of abstraction which usually accompanies the introduction of e. The appearance of e is followed by an introduction of exponential functions with applications to continuous compounding of interest situations. The contents of Chapter 1 are sometimes treated in precalculus courses. If a class has had such a course, then Chapter 1 can either be quickly surveyed or omitted entirely.

Chapter 2 begins with stage two in the development of the derivative. Several concepts, in different settings, which exhibit the derivative structure (tangent line to a curve at a point, marginal revenue, marginal cost, instantaneous velocity) are introduced. With the stage thus set, the derivative is seen to emerge as a mathematical concept having a structure common to this variety of situations. Having established the importance of the derivative, tools to facilitate the calculation of derivatives are developed. These tools are then applied to a variety of situations which exhibit the derivative structure (Section 17).

With the derivative machinery established in Chapter 2, it seemed desirable to apply it as soon as possible to situations whose significance is immediately apparent. It seemed to me that optimization problems, as opposed to graph-sketching techniques, best fulfill this requirement. Thus Chapter 3 is given over to a discussion of extreme values of functions with applications to the profit-maximization behavior of a firm, tax-maximization problems, and optimal storage time problems. To maximize flexibility, Chapter 3 does not presuppose graph-sketching techniques, material which is discussed in Chapter 4. To graph the functions which arise in the application of optimization methods in Chapter 3 to the profit-maximization behavior of the firm, it suffices to plot a few points and connect them with a smooth curve. On the other hand, the graph-sketching techniques of Chapter 4 can be taken up before the application of the optimization methods of Chapter 3, if one wishes to do so.

It has been my experience that students find the "LUB–GLB" approach to the definite integral much easier to understand than the usual "limit of sums" approach. This is not surprising; the "limit of sums" approach involves a peculiar and quite complicated passage to the limit, which is avoided in the "LUB–GLB" approach. For this reason the definite integral is first presented in Chapter 5 via the "LUB–GLB" approach and then considered again from the "limit of sums" point of view. While this is my personal preference, an instructor can, if he wishes, begin with the "limit of sums" approach or present only one of these approaches. As with the derivative concept, the formal introduction of the definite integral is prefaced by a discussion of concepts from different settings which exhibit the definite integral structure (area and work were chosen because they seemed most natural and most simple). With the stage thus set, the definite integral is then introduced as a mathematical concept having the structure common to these situations. Having established the importance of the definite integral, tools to facilitate the calculation of definite integrals are developed. These tools are then applied to a variety of situations which exhibit the definite integral structure (Section 29).

My objective in Chapter 6 is to present multivariable functions and their applications and get to the concept of partial derivative as quickly and as simply as possible. A number of concepts which exhibit the partial derivative structure are pointed out. To whet the appetite, the chapter concludes with a brief look at multivariable optimization problems.

Again it is my pleasure to acknowledge the contributions of Carol Beal, Arthur B. Evans, and the rest of the staff at Xerox College Publishing in the preparation of this book, and to thank the reviewers (especially Professor Donald Burleson of Middlesex Community College, and Professor Paul T. Banks of Boston College) for their sensitive and helpful comments.

W. J. A.

Contents

CHAPTER DEPENDENCES

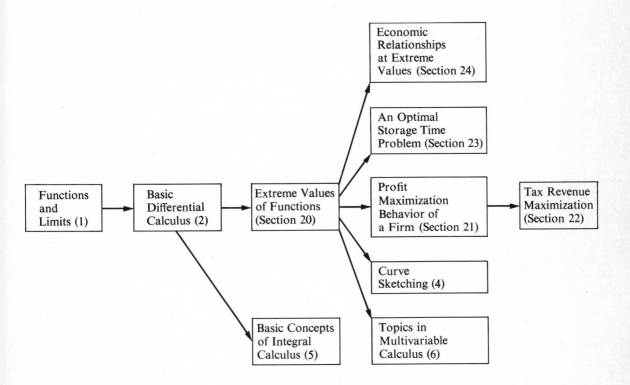

1

Functions and limits

1. Functions of one variable

Many relationships in the biological, business, physical, and social sciences exhibit the following structure. There are two variables, x and y, let us say, and a scheme or rule or relation by means of which there is assigned to each value of x exactly one value of y. Such a structure is called a *function*. We also say that y is a *function of x*. Thus, for example, the rule

$$y = 2x^2 + 2$$

defines y as a function of x. Given any value for x, we can determine the corresponding value for y; thus if $x = 0$, $y = 2$; if $x = 1$, $y = 4$; if $x = -2$, $y = 10$, and so on. The rule

$$y = \frac{1}{x}$$

defines y as a function of x. If $x = 2$, $y = \frac{1}{2}$; if $x = -3$, $y = -\frac{1}{3}$, and so on.

If y is a function of x, then x is called the *independent variable* and y is called the *dependent variable*. The set of all values which can be assumed by the independent variable is called the *domain of definition* (or *domain*) of the variable and the function. If no explicit mention is made of the domain of the function, then it is understood to consist of all real numbers for which its rule is meaningful. Thus the domain of definition of $y = 2x^2 + 2$ is the set of real numbers. The domain of definition of $y = 1/x$ is the set of all real numbers except zero. Zero must be excluded since division by zero is not defined.

1

It is common practice to use letters such as f, g, h, F, G, ϕ (Greek "phi"), and ψ (Greek "psi"), to designate either the rule of the function or the function itself. In terms of this notation, the symbol $f(x)$, read "f of x" or "f at x," is used to represent the value assigned by f to x. For example,

$$y = 2x^2 + 2$$

would be expressed in terms of this notation by

$$f(x) = 2x^2 + 2.$$

Thus $f(x)$ is another symbol for the dependent variable y. The symbol $f(0)$ stands for the value of the dependent variable, or the value of the function, when $x = 0$. Hence for $f(x) = 2x^2 + 2$, $f(0) = 2(0)^2 + 2 = 2$.

$f(1) = 2(1)^2 + 2 = 4$; the value of function f when $x = 1$ is 4.
$f(3) = 2(3)^2 + 2 = 20$; the value of function f when $x = 3$ is 20.
$f(a) = 2(a)^2 + 2$; the value of function f when $x = a$ is $2a^2 + 2$.

In the case of $g(x) = 1/x$, we say that $g(0)$ does not exist, or is not defined, since 0 is not in the domain of x.

Two functions f and g are said to be *equal* if

1. they have the same domain of definition;
2. their rules assign the same value to each number in their common domain of definition.

Thus f and g defined by

$$f(x) = x + 3$$
$$g(t) = t + 3$$

are equal. The domain of each is the set of real numbers and their rules assign the same value to each given real number. Thus we see that it makes no difference what symbol is used for the independent variable of a function.

The functions f and g defined by

$$f(x) = 2x^2 + 2$$
$$g(x) = 2x^2 + 2, \qquad -1 < x < 1$$

are not equal since their domains of definition are different. That is, the domain of definition of f is the set of real numbers, but the domain of definition of g consists of only those real numbers between -1 and 1.

The following examples further illustrate the concept of function.

EXAMPLE 1.   Let f denote the function defined on the set of real numbers by

$$f(x) = \begin{cases} 2x + 1, & x \le 1 \\ x - 2, & x > 1. \end{cases}$$

This is one function—one rule and one domain of definition—whose rule is described by two algebraic expressions. The expression to be used depends on whether the number in the domain is less than or equal to 1, or greater than 1. Thus $f(1) = 2(1) + 1 = 3$, $f(-2) = 2(-2) + 1 = -3$, but $f(4) = 4 - 2 = 2$.

EXAMPLE 2. Let f denote the function defined on the set of real numbers which assigns to each real number the value 2; that is,

$$f(x) = 2.$$

Thus $f(0) = 2$, $f(-3) = 2$, $f(\sqrt{3}) = 2$, $f(100) = 2$, and so on. For obvious reasons this function is called a *constant function*.

EXAMPLE 3. In the mathematics of finance it is shown that if $1000 is deposited in a bank paying interest at a rate of 6% compounded annually, then the amount on deposit after n years is $1000(1.06)^n$ (see Appendix 2). Thus the capital-accumulation function which describes the capital accumulated after t years in this situation is defined by the following function $C(t)$.

$$C(t) = \begin{cases} 1000, & 0 \leq t < 1 \\ 1000(1.06), & 1 \leq t < 2 \\ 1000(1.06)^2, & 2 \leq t < 3 \\ 1000(1.06)^3, & 3 \leq t < 4 \\ \quad \vdots \\ 1000(1.06)^n & n \leq t < n+1 \\ \quad \vdots \end{cases}$$

$$A = P\left(1 + R/100\right)^N$$

Thus, for example, $C(1) = \$1060$ and $C(3) = 1000(1.06)^3 = \$1190.38$.

EXAMPLE 4. From the tax-rate schedule for the 1973 New York City income tax given in Table 1, we obtain the income-tax function $T(x)$ defined as shown here.

$$T(x) = \begin{cases} (0.007)x, & 0 < x \leq 1,000 \\ 7 + 0.011(x - 1,000), & 1,000 < x \leq 3,000 \\ 29 + 0.014(x - 3,000), & 3,000 < x \leq 6,000 \\ 71 + 0.018(x - 6,000), & 6,000 < x \leq 10,000 \\ 143 + 0.021(x - 10,000), & 10,000 < x \leq 15,000 \\ 248 + 0.025(x - 15,000), & 15,000 < x \leq 20,000 \\ 373 + 0.028(x - 20,000), & 20,000 < x \leq 25,000 \\ 513 + 0.032(x - 25,000), & 25,000 < x \leq 30,000 \\ 673 + 0.035(x - 30,000), & 30,000 < x \end{cases}$$

If taxable income is		The tax to be paid is
not over $1,000		0.7% (0.007) of taxable income

over	but not over				
$ 1,000	$ 3,000	$ 7	plus	1.1% (0.011)	of excess over $ 1,000
3,000	6,000	29	plus	1.4% (0.014)	of excess over 3,000
6,000	10,000	71	plus	1.8% (0.018)	of excess over 6,000
10,000	15,000	143	plus	2.1% (0.021)	of excess over 10,000
15,000	20,000	248	plus	2.5% (0.025)	of excess over 15,000
20,000	25,000	373	plus	2.8% (0.028)	of excess over 20,000
25,000	30,000	513	plus	3.2% (0.032)	of excess over 25,000
30,000		673	plus	3.5% (0.035)	of excess over 30,000

Table 1

Thus if taxable income is $9000, the tax to be paid is $T(9000) = 71 + 0.018(3000) = \$125$; if taxable income is $12,000, the tax to be paid is $T(12,000) = 143 + 0.021(2000) = \$185$.

EXAMPLE 5. *Demand for a commodity as a function of price.* A relation studied in economics is the functional relationship between the price of a commodity and the amount of the commodity required by the market when other factors that affect demand— such as income, tastes, prices of substitute commodities—are held constant. If D is the quantity demanded by the market in a unit of time and p is the price of a unit of the commodity, then the general relationship between them can be expressed in function notation by

$$D = f(p).$$

Thus suppose

$$D = -5p + 30$$

is the demand function which expresses the market demand for pork in millions of pounds per month as a function of price of pork in dollars per pound. The domain of definition of p is understood to be all numbers for which $D = -5p + 30$ is economically as well as mathematically meaningful. (Such is the case in general.) In this example, households will purchase 25 million pounds of pork per month if the price is $1.00 per pound:

$$D = -5(1) + 30 = 25.$$

They will purchase 22.5 million pounds of pork per month if the price is $1.50 per pound:

$$D = -5(1.5) + 30 = 22.5.$$

In many instances it is assumed that the demand function is a linear function of the form

$$D = -ap + b$$

where a and b are positive constants. In our illustration $a = 5$ and $b = 30$.

EXAMPLE 6. *Supply of a commodity as a function of price.* Another relationship studied in economics is the one between the supply S of a commodity (the amount of the commodity that firms wish to sell per unit of time) and its market price p, under the assumption that everything that affects the supply of the commodity other than its price is held constant. The general relationship between S and p can be expressed in function notation by

$$S = f(p).$$

Thus suppose

$$S = 2p - 10$$

is the supply function which expresses the supply of wheat in millions of pounds per month as a function of the market price of wheat in cents per pound. In this example, if the market price is 30¢ per pound, the quantity that will be supplied to the market is 50 million pounds per month:

$$S = 2(30) - 10 = 50.$$

If the market price is 25¢ per pound, the quantity that will be supplied to the market is 40 million pounds per month:

$$S = 2(25) - 10 = 40.$$

In many instances it is assumed that the supply function is a linear function of the form

$$S = ap - b$$

where a and b are positive constants. In our example $a = 2$ and $b = 10$.

EXAMPLE 7. *Pareto's income-distribution functions.* The economist Vilfredo Pareto* advanced the class of functions

$$y = \frac{a}{x^m}$$

as a description of the number of incomes y which exceed the amount x. a and m are population constants which depend on the population, with the value of m usually being found to be near 1.5.

To illustrate, let us suppose that the Pareto income-distribution function for a

---

* *Cours d'économie politique* (Lausanne, 1897), vol. 2, book 3, Chapter 1.

certain population is

$$y = \frac{10^{10}}{x^{3/2}}$$

where the monetary units are, let us say, dollars. Then the number of incomes over $10,000 is

$$y = \frac{10^{10}}{(10^4)^{3/2}} = \frac{10^{10}}{10^6} = 10,000.$$

The number of incomes over $1,000,000 is

$$y = \frac{10^{10}}{(10^6)^{3/2}} = \frac{10^{10}}{10^9} = 10.$$

The number of incomes between $10,000 and $1,000,000, including $1,000,000, is $10,000 - 10 = 9990$.

EXAMPLE 8. *Cost functions.* Consider a firm which is producing a single, uniform commodity. The function

$$c = f(x)$$

which describes the total cost c for the production of x units per unit time, is called the total cost function of the firm. For example, suppose

$$c = \tfrac{1}{5}x^2 + 10x + 200$$

is the cost function for a sugar refinery which expresses the cost c in dollars per week as a function of the output x of tons of sugar per week. Thus, in this situation, the cost of producing 5 tons a week would be $255; the cost of producing 10 tons of sugar a week would be $320.

For any output x, the ratio of total cost c to output x defines average cost, or cost per unit output. Denoting average cost by $\bar{c}$, we have

$$\bar{c} = \frac{c}{x} = \frac{f(x)}{x}.$$

For our sugar refinery the average cost function is

$$\bar{c} = \frac{1}{5}x + 10 + \frac{200}{x}.$$

The average cost per ton for a weekly output of 5 tons is $51.

Exercises

1. For $f(x) = 2x^2 + 3x - 4$, find $f(0)$, $f(1)$, $f(-2)$, and $f(-1)$.
2. For $f(x) = 3$, find $f(0)$, $f(-4)$, $f(3)$, and $f(1)$.

3. For $f(x) = 3x - 2$, defined on $D = \{-1, 1, 2, 3\}$, find $f(-1)$, $f(0)$, $f(1)$, $f(2)$, and $f(3)$.

4. For $f(x) = \begin{cases} 2, & x \geq 3 \\ 1, & x < 3 \end{cases}$, find $f(0)$, $f(1)$, $f(\frac{1}{2})$, $f(3)$, $f(4)$, $f(7)$, and $f(24)$.

5. For $f(x) = \begin{cases} 2, & x \geq 3 \\ x + 1, & x < 3 \end{cases}$, find $f(0)$, $f(1)$, $f(\frac{1}{2})$, $f(3)$, $f(4)$, $f(7)$, and $f(24)$.

6. For $f(x) = \begin{cases} x^2, & x \geq 0 \\ 3, & x < 0 \end{cases}$, find $f(0)$, $f(\frac{1}{2})$, $f(1)$, $f(2)$, $f(3)$, $f(-\frac{1}{2})$, $f(-1)$, $f(-\frac{3}{2})$, and $f(-2)$.

7. For $f(x) = \begin{cases} x^2, & x \neq 2 \\ 1, & x = 2 \end{cases}$, find $f(-2)$, $f(-1)$, $f(0)$, $f(\frac{1}{2})$, $f(1)$, $f(2)$, and $f(3)$.

8. Are the functions $f(x) = \dfrac{x^2 - 4}{x - 2}$ and $g(x) = x + 2$ equal? Explain.

9. For $f(x) = 16x^2$, determine $g(x) = \dfrac{f(x) - f(1)}{x - 1}$. What is the domain of definition of $g(x)$?

10. For $f(t) = 2t^2 + 1$, determine $g(t) = \dfrac{f(t) - f(2)}{t - 2}$. What is the domain of definition of $g(t)$?

11. For $f(x) = 600x - 5x^2$, determine $g(x) = \dfrac{f(x) - f(50)}{x - 50}$.

12. For $f(x) = \dfrac{1}{2}x^2 + x + 1$, determine $g(x) = \dfrac{f(x) - f(2)}{x - 2}$.

13. For $f(p) = 120 - \dfrac{1}{5}p$, determine $g(p) = \dfrac{f(p) - f(50)}{p - 50}$.

14. For $f(x) = 2$, find $g(x) = \dfrac{f(x) - f(1)}{x - 1}$ and $h(x) = \dfrac{f(x) - f(a)}{x - a}$.

15. For $f(x) = 100 - x^2$, determine $g(x) = \dfrac{f(x) - f(3)}{x - 3}$.

16. For $f(x) = \dfrac{1}{5}x^2 + 3x + 1$, determine $g(x) = \dfrac{f(x) - f(10)}{x - 10}$.

17. For $f(x) = \begin{cases} 2, & x \geq 1 \\ 1, & x < 1 \end{cases}$ and $g(x) = \begin{cases} 1, & x \geq 1 \\ 2, & x < 1 \end{cases}$, find $h(x) = f(x) + g(x)$.

18. For $f(x) = \begin{cases} 2, & x \geq 3 \\ x + 1, & x < 3 \end{cases}$ and $g(x) = \begin{cases} x^2, & x \geq 3 \\ 1 - x, & x < 3 \end{cases}$, find $h(x) = f(x) + g(x)$.

19. A taxi charges 60¢ for the first one-fifth of a mile and 10¢ for each additional one-fifth of a mile. Express cost as a function of distance traveled.

20. Determine the income-tax function T which expresses tax to be paid as a function of taxable income for the 1973 New York State tax schedule given on page 8.

Taxable Income		Tax to be Paid			
over	but not over				
$ 0	$ 1,000		2%	of taxable income	
1,000	3,000	$ 20	plus 3%	of excess over	$ 1,000
3,000	5,000	80	plus 4%	of excess over	3,000
5,000	7,000	160	plus 5%	of excess over	5,000
7,000	9,000	260	plus 6%	of excess over	7,000
9,000	11,000	380	plus 7%	of excess over	9,000
11,000	13,000	520	plus 8%	of excess over	11,000
13,000	15,000	680	plus 9%	of excess over	13,000
15,000	17,000	860	plus 10%	of excess over	15,000
17,000	19,000	1,060	plus 11%	of excess over	17,000
19,000	21,000	1,280	plus 12%	of excess over	19,000
21,000	23,000	1,520	plus 13%	of excess over	21,000
23,000	25,000	1,780	plus 14%	of excess over	23,000
25,000		2,060	plus 15%	of excess over	25,000

21. A lamp manufacturer has fixed expenses of $1000 per week. His production cost is $5 per lamp. It is estimated that if the selling price of the lamp is x dollars, then $450 - 10x$ lamps would be sold per week. Express weekly income as a function of selling price.

22. A manufacturer is planning to market drums of oil. Each drum is to have a volume of 12 cubic feet. The material needed to make the drums costs 10¢ per square foot. Express the cost of material as a function of the radius of the drum.

2. Graphs of functions

By the *graph* of the function $y = f(x)$, we mean the collection of all points with coordinates $(x, f(x))$, where x takes on only those values in its domain of definition. The graph of a function provides us with a visual image of the function and reveals the behavior of the function at a glance.

EXAMPLE 9. Sketch the graph of the linear function $y = 2x + 1$ defined on the set of real numbers.

Solution. Since the graph of a linear function defined on the set of real numbers is a line, it suffices to plot two points to determine its graph. If $x = 0$, $y = 1$; if $x = 1$, $y = 3$. Thus $(0, 1)$ and $(1, 3)$ are two points on the graph, which is shown in Figure 1.

EXAMPLE 10. Sketch the graph of $y = 2x + 1$ defined on $D = \{1, 2, 3\}$.

Solution. Since x can only take on the values 1, 2, and 3, the graph of this function consists of the three points with coordinates $(1, 3)$, $(2, 5)$, and $(3, 7)$, as shown in Figure 2.

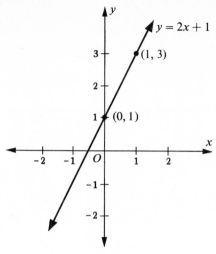

Figure 1

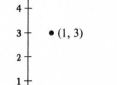

Figure 2

EXAMPLE 11. Sketch the graph of $f(x) = 2$ defined on the set of real numbers.

Solution. The graph of $f(x) = 2$, or $y = 2$, is the horizontal line two units above the x-axis and parallel to it (see Figure 3).

EXAMPLE 12. Sketch the graph of the function f defined on the set of real numbers by

$$f(x) = \begin{cases} 1, & x \geq 0 \\ -1, & x < 0 \end{cases}.$$

Solution. For $x \geq 0, f(x) = 1$; thus we obtain a ray one unit above the x-axis with end point $(0, 1)$. For $x < 0, f(x) = -1$, thus yielding a half-line one unit below the x-axis. The end point $(0, -1)$ is not on the graph since $f(0)$ is 1. The graph of f is shown in Figure 4.

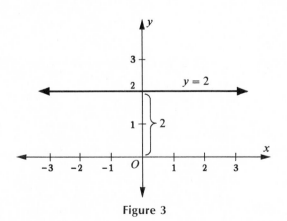

Figure 3

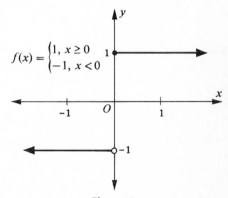

Figure 4

EXAMPLE 13. Sketch the graph of $h(x) = \dfrac{x^2 - 9}{x - 3}$.

Solution. First let us observe that the domain of definition of this function is the set of real numbers except 3. $h(3)$ is not defined since the substitution of 3 for x in the algebraic expression defining h yields 0 in the denominator, and division by 0 is not defined. We can simplify $h(x)$ by factoring. For $x \neq 3$ we have

$$h(x) = \frac{x^2 - 9}{x - 3} = \frac{(x - 3)(x + 3)}{(x - 3)} = x + 3.$$

Thus the graph of $h(x) = x + 3$, with $x \neq 3$, consists of all points on the line $y = x + 3$ with the exception of $(3, 6)$ (see Figure 5).

EXAMPLE 14. Sketch the graph of

$$m(x) = \begin{cases} \dfrac{x^2 - 9}{x - 3}, & x \neq 3 \\ 5, & x = 3 \end{cases}$$

Solution. Let us observe that $m(x)$ is the same as $h(x)$, defined in Example 13, for $x \neq 3$. While $h(3)$ is not defined, $m(3) = 5$. Thus the graph of $m(x)$ consists of the point $(3, 5)$ and all points on the line $y = x + 3$ with the exception of $(3, 6)$ (see Figure 6).

A crude, but often effective, approach to graphing certain nonlinear functions is to let the independent variable take on a number of values, calculate the corresponding values of the dependent variable, tabulate the results, plot the corresponding points, and join them with a smooth curve. The examples just considered serve to indicate that this technique must be used with discretion and that indis-

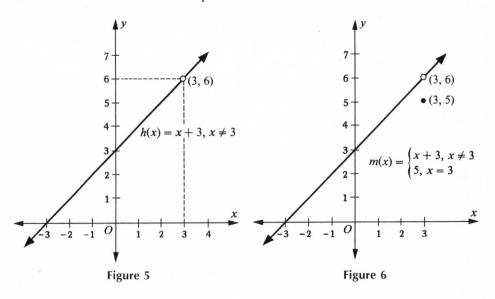

Figure 5   Figure 6

criminate use could lead to incorrect results. In Chapter 4, Curve Sketching, we will see how tools forged from the calculus aid us in graphing functions.

EXAMPLE 15.   Sketch the graph of $f(x) = x^2$.

Solution.   This function is defined on the set of real numbers and its graph has no jumps or breaks (see Example 29, Section 5). From $f(x) = x^2$ we obtain the following table of values.

x	-3	-2	-1	0	1	2	3
$f(x)$	9	4	1	0	1	4	9

Plotting these points and joining them with a smooth curve yields the graph shown in Figure 7.

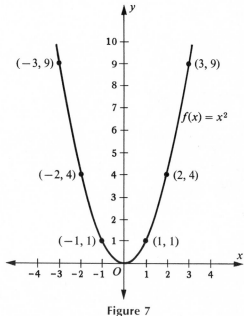

Figure 7

EXAMPLE 16.   Sketch the graph of $f(x) = 1/x$.

Solution.   This function is defined for all real numbers except zero. Its graph has no jumps or breaks except at zero (see Example 30, Section 5). From $f(x) = 1/x$ we obtain the following table of values.

x	-1000	-100	-10	-1	$-\frac{1}{10}$	$-\frac{1}{100}$	$-\frac{1}{1000}$	$\frac{1}{1000}$	$\frac{1}{100}$	$\frac{1}{10}$	1	100	1000
$f(x)$	$-\frac{1}{1000}$	$-\frac{1}{100}$	$-\frac{1}{10}$	-1	-10	-100	-1000	1000	100	10	1	$\frac{1}{100}$	$\frac{1}{1000}$

From this table we see that as x takes on positive values which are decreasing and closer and closer to 0, that is, as x approaches zero $(1, \frac{1}{10}, \frac{1}{100}, \frac{1}{1000})$, the corresponding values of $f(x) = 1/x$ increase without bound $(1, 10, 100, 1000)$. This behavior is shown in Figure 8a. As x takes on positive values which are increasing without bound $(1, 10, 100, 1000)$, corresponding values of $f(x) = 1/x$ approach 0 in a positive sense $(1, \frac{1}{10}, \frac{1}{100}, \frac{1}{1000})$. See Figure 8b. As x approaches zero in a negative sense $(-1, -\frac{1}{10}, -\frac{1}{100}, -\frac{1}{1000})$, $f(x) = 1/x$ decreases without bound $(-1, -10, -100, -1000)$. See Figure 8c. As x takes on negative values which are decreasing without bound $(-1, -10, -100, -1000)$, $f(x) = 1/x$ approaches zero in a negative sense $(-1, -\frac{1}{10}, -\frac{1}{100}, -\frac{1}{1000})$. The graph of $f(x) = 1/x$ is shown in Figure 8d.

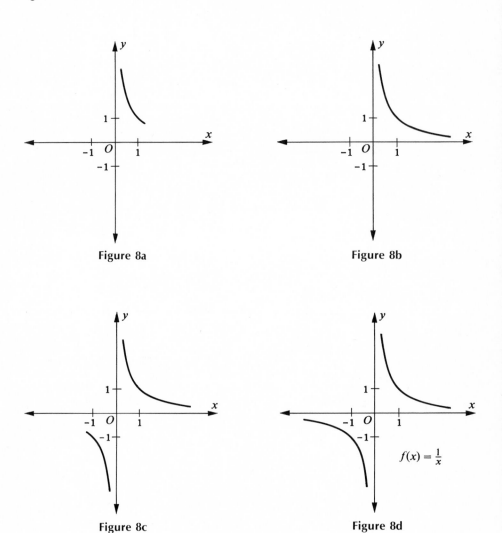

Figure 8a

Figure 8b

Figure 8c

Figure 8d

Exercises

Sketch the graphs of the following functions.

23. $f(x) = 3$

24. $f(x) = 3x - 2$, defined on $D = \{-1, 1, 2, 3\}$

25. $f(x) = x + 1$, where $0 \leq x \leq 2$

26. $f(x) = \begin{cases} 2, & x \geq 3 \\ 1, & x < 3 \end{cases}$

27. $f(x) = \begin{cases} 2, & x \geq 3 \\ x + 1, & x < 3 \end{cases}$

28. $f(x) = \begin{cases} x^2, & x \geq 0 \\ 3, & x < 0 \end{cases}$

29. $f(x) = \begin{cases} x^2, & x \neq 2 \\ 1, & x = 2 \end{cases}$

30. $f(x) = \dfrac{x^2 - 4}{x - 2}$ and $g(x) = x + 2$

31. $f(x) = \begin{cases} x, & x \geq 0 \\ -x, & x < 0 \end{cases}$

32. $f(x) = \begin{cases} x, & x \leq 1 \\ x + 1, & x > 1 \end{cases}$

33. $f(x) = x^2 + 1$

34. $f(x) = -x^2 + 1$

35. $f(x) = 1/x^2$

36. $f(x) = -(1/x^2)$

37. If $1000 is deposited in an account paying interest at a rate of 6% compounded annually, then the amount on deposit t years later, where $0 \leq t < 3$, is described by the capital-accumulation function below.

$$C(t) = \begin{cases} 1000, & 0 \leq t < 1 \\ 1000(1.06), & 1 \leq t < 2 \\ 1000(1.06)^2, & 2 \leq t < 3 \end{cases}$$

Sketch the graph of this capital-accumulation function.

38. Sketch the graph of the tax function discussed in Example 4 of Section 1.

3. Limits of functions

A question which often arises is this: As x takes on values closer and closer to a given fixed value a, how does $f(x)$ behave? Specifically, does $f(x)$ remain constant or get closer and closer to a single value? If as x takes on values closer and closer to a value a, $f(x)$ gets closer and closer to a single value L, or remains constant at L, then we define L to be the *limit of function f as x approaches a*. This is written

$$\lim_{x \to a} f(x) = L \quad \text{or} \quad f(x) \to L \text{ as } x \to a.$$

This notation is read, "the limit of $f(x)$ is L as x approaches a," or, "$f(x)$ approaches L as x approaches a."

EXAMPLE 17. If f is defined by $f(x) = 2x + 1$, then $\lim_{x \to 2} f(x) = 5$. That is, as x takes on values closer and closer to 2, $f(x) = 2x + 1$ gets closer and closer to a single number, 5. $f(x) = 2x + 1$ can be made as close as we wish to 5 by taking x sufficiently close to 2. More precisely, if you were to say that you want $2x + 1$ to be within d units of 5, then we would be able to specify how close to 2 the value of x should be to have $2x + 1$ within d units of 5. If d is as shown in Figure 9, then x would have to be within h units of 2 to have $2x + 1$ within d units of 5.

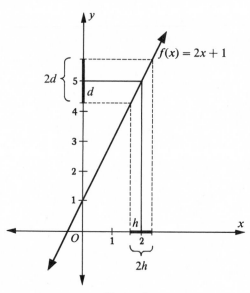

Figure 9

EXAMPLE 18. If f is defined on the set of real numbers by

$$f(x) = \begin{cases} 1, & x \geq 0 \\ -1, & x < 0 \end{cases}$$

then $\lim_{x \to 0} f(x)$ does not exist. From the graph of f (Figure 10) we see that as x takes on values closer and closer to 0, $f(x)$ does not get closer and closer to a single

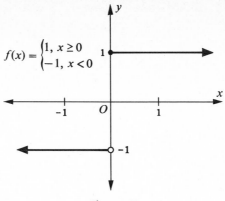

$$f(x) = \begin{cases} 1, & x \geq 0 \\ -1, & x < 0 \end{cases}$$

Figure 10

value or remain constant. Instead $f(x)$ clusters about two values, -1 and 1. This leads us to say that $\lim\limits_{x \to 0} f(x)$ does not exist.

EXAMPLE 19.   If $h(x) = \dfrac{x^2 - 9}{x - 3}$, then $\lim\limits_{x \to 3} h(x) = 6$. For $x \neq 3$,

$$h(x) = \frac{x^2 - 9}{x - 3} = \frac{(x - 3)(x + 3)}{(x - 3)} = x + 3.$$

The graph of $h(x)$—consisting of all points on the line $y = x + 3$ with the exception of $(3, 6)$—is discussed in Example 13 (Section 2) and shown here in Figure 11. As x takes on values closer and closer to 3, $h(x) = x + 3$, $x \neq 3$, gets closer and closer

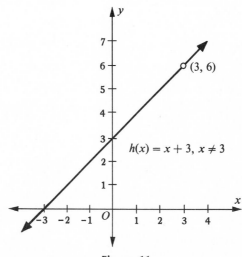

Figure 11

to 6. We can make $x + 3$ as close as we wish to 6 by taking x sufficiently close to 3. Thus by definition $\underset{x \to 3}{\text{limit}}\, h(x) = 6$.

E X A M P L E   20.   If

$$m(x) = \begin{cases} \dfrac{x^2 - 9}{x - 3}, & x \neq 3 \\ 5, & x = 3 \end{cases}$$

then $\underset{x \to 3}{\text{limit}}\, m(x) = 6$. The graph of $m(x)$—consisting of the point $(3, 5)$ and all points on the line $y = x + 3$ except $(3, 6)$—is discussed in Example 14 (Section 2) and shown here in Figure 12. As x takes on values closer and closer to 3, $m(x)$ gets closer and closer to 6. We can make $m(x)$ as close to 6 as we wish by taking x sufficiently close to 3.

Examples 19 and 20 serve to illustrate a point which is sometimes overlooked, namely, that the behavior of a function at a given value has nothing to do with whether or not the function has a limit with respect to that value and what the limit is. The question basic to the existence of a limit is this: How do the values of the function behave as x takes on values closer and closer to the given value? Irrelevant to this concern is the behavior of the function at the given value. From Example 19 we see that the limit of $h(x)$ as x approaches 3 exists even though $h(3)$ is not defined. From Example 20 we see that the way in which a function is defined at a value (3 in this case) and the behavior of the function as x is taken closer and closer to that value can be quite different: $m(3) = 5$, which $\underset{x \to 3}{\text{limit}}\, m(x) = 6$.

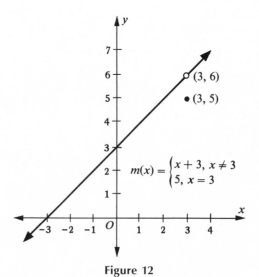

Figure 12

E X A M P L E   21.   If $f(x) = 2$, for all real numbers, then $\underset{x \to 1}{\text{limit}} f(x) = 2$. As x takes on values closer and closer to $1, f(x) = 2$ remains constant at 2. Thus $\underset{x \to 1}{\text{limit}} f(x) = 2$. More generally, if a is any real number, then $\underset{x \to a}{\text{limit}} f(x) = 2$.

   Still more generally, if $f(x) = c$ is any constant function and a is any real number, then

$$\underset{x \to a}{\text{limit}} f(x) = c.$$

E X A M P L E   22.   For $f(x) = 16x^2$, determine, if it exists,

$$\underset{x \to 1}{\text{limit}} \frac{f(x) - f(1)}{x - 1}.$$

Solution. $f(x) = 16x^2, f(1) = 16$; thus

$$\frac{f(x) - f(1)}{x - 1} = \frac{16x^2 - 16}{x - 1}$$

$$= \frac{16(x^2 - 1)}{x - 1} = \frac{16(x - 1)(x + 1)}{x - 1}$$

$$= 16(x + 1) = 16x + 16$$

where $x \neq 1$. Thus

$$\underset{x \to 1}{\text{limit}} \frac{f(x) - f(1)}{x - 1} = \underset{x \to 1}{\text{limit}} (16x + 16) = 32.$$

As x takes on values closer and closer to 1, $16x + 16$ gets closer and closer to 32.

Exercises

Use the definition of limit to determine if the limits of the following functions exist. When a limit exists, find its value. When a limit does not exist, explain why.

39. $\underset{x \to 2}{\text{limit}} f(x)$, where $f(x) = 2x$

40. $\underset{x \to 2}{\text{limit}} f(x)$, where $f(x) = x^2$

41. $\underset{x \to -2}{\text{limit}} f(x)$, where $f(x) = 3x^2 - 2$

42. $\underset{x \to 3}{\text{limit}} f(x)$, where $f(x) = 2x - 3$

43. $\underset{x \to 4}{\text{limit}} f(x)$, where $f(x) = 2x^2 + 1$

44. $\underset{x \to -3}{\text{limit}} f(x)$, where $f(x) = 3x + 4$

45. $\underset{x \to 2}{\text{limit}} f(x)$, where $f(x) = x^2 + 3x$

46. $\underset{x \to -2}{\text{limit}} f(x)$, where $f(x) = 2x^2 - x + 1$

47. $\displaystyle\lim_{x \to 2} f(x)$, where $f(x) = 2x, \ x \neq 2$

48. $\displaystyle\lim_{x \to 2} f(x)$, where $f(x) = \begin{cases} x^2, & x \neq 2 \\ 1, & x = 2 \end{cases}$

49. $\displaystyle\lim_{x \to 2} f(x)$, where $f(x) = \dfrac{x^2 - 4}{x - 2}$

50. $\displaystyle\lim_{x \to 0} f(x)$, where $f(x) = \begin{cases} x + 1, & x \geq 0 \\ -x + 1, & x < 0 \end{cases}$

51. $\displaystyle\lim_{x \to -4} f(x)$, where $f(x) = \dfrac{x^2 - 16}{x + 4}$

52. $\displaystyle\lim_{x \to 3} f(x)$, where $f(x) = \begin{cases} 2, & x \geq 3 \\ 1, & x < 3 \end{cases}$

53. $\displaystyle\lim_{x \to 2} f(x)$, where $f(x) = \begin{cases} 2, & x \geq 3 \\ 1, & x < 3 \end{cases}$

54. $\displaystyle\lim_{x \to 3} f(x)$, where $f(x) = \begin{cases} 2, & x \geq 3 \\ x + 1, & x < 3 \end{cases}$

55. $\displaystyle\lim_{x \to 4} f(x)$, where $f(x) = \begin{cases} 2x + 1, & x \neq 4 \\ -3, & x = 4 \end{cases}$

56. $\displaystyle\lim_{x \to 0} f(x)$, where $f(x) = \begin{cases} x^2, & x \geq 0 \\ 3, & x < 0 \end{cases}$

57. $\displaystyle\lim_{x \to 0} f(x)$, where $f(x) = \begin{cases} 2, & x \geq 3 \\ 1, & x < 3 \end{cases}$

4. Fundamental limit theorems

Since the use of the definition of limit to determine specific limits is only possible in certain situations, it is quite natural to try to construct tools which would enable us to determine limits in a relatively simple way. Perhaps the most natural approach to this problem is to regard a function as being made up of component parts, put together by the operations of addition, multiplication, and so on, and then to examine the limit behavior of the component parts. If the component parts have limits, then it would seem likely that the function itself has a limit and that this limit can be determined from the limits of the component parts.

Such an approach to the problem of determining limits is not only natural but also fruitful. The following theorems can be established on the basis of the definition of limit. When applicable, they are powerful tools.

Limit of a Sum. Suppose $h(x)$ can be expressed as a sum of functions $f(x)$ and $g(x)$, that is, $h(x) = f(x) + g(x)$, and $\displaystyle\lim_{x \to a} f(x)$ and $\displaystyle\lim_{x \to a} g(x)$ exist. Then

$$\lim_{x \to a} h(x) = \lim_{x \to a} f(x) + \lim_{x \to a} g(x).$$

In everyday language this theorem is sometimes stated, the limit of a sum is the sum of the limits. This of course presupposes that the members of the sum have limits. More generally, this theorem holds for a sum of two or more functions.

Limit of a Difference. Suppose $h(x)$ can be expressed as a difference of functions $f(x)$ and $g(x)$, that is, $h(x) = f(x) - g(x)$, and $\lim_{x \to a} f(x)$ and $\lim_{x \to a} g(x)$ exist. Then

$$\lim_{x \to a} h(x) = \lim_{x \to a} f(x) - \lim_{x \to a} g(x).$$

In colloquial language, the limit of a difference is the difference of the limits. More generally, this theorem holds for a difference of two or more functions.

Limit of a Product. Suppose $h(x)$ can be expressed as a product of functions $f(x)$ and $g(x)$, that is, $h(x) = f(x) \cdot g(x)$, and $\lim_{x \to a} f(x)$ and $\lim_{x \to a} g(x)$ exist. Then

$$\lim_{x \to a} h(x) = [\lim_{x \to a} f(x)] \cdot [\lim_{x \to a} g(x)].$$

In colloquial language, the limit of a product is the product of the limits. More generally, this theorem holds for a product of two or more functions.

Limit of a Quotient. Suppose $h(x)$ can be expressed as a quotient of functions $f(x)$ and $g(x)$, that is, $h(x) = f(x)/g(x)$, and $\lim_{x \to a} f(x)$ exists and $\lim_{x \to a} g(x)$ exists and is not zero. Then

$$\lim_{x \to a} h(x) = \frac{\lim_{x \to a} f(x)}{\lim_{x \to a} g(x)}.$$

In colloquial language, the limit of a quotient is the quotient of the limits.

E X A M P L E   23.   Determine, if it exists, $\lim_{x \to 2} 3x^2$.

Solution. $h(x) = 3x^2$ can be expressed as the product of the constant function 3, the function x, and the function x. Each of these functions has a limit with respect to 2.

$$\lim_{x \to 2} 3x^2 = [\lim_{x \to 2} 3] \cdot [\lim_{x \to 2} x] \cdot [\lim_{x \to 2} x] = 3 \cdot 2 \cdot 2 = 12$$

E X A M P L E   24.   Determine, if it exists, $\lim_{x \to 2} (3x^2 + 4)$.

Solution. $h(x) = 3x^2 + 4$ can be viewed as the sum of the function $3x^2$ and the constant function 4, each of which has a limit as x approaches 2. From Example 23, $\lim_{x \to 2} 3x^2 = 12$. Thus by the limit theorem for sums,

$$\lim_{x \to 2} (3x^2 + 4) = \lim_{x \to 2} 3x^2 + \lim_{x \to 2} 4 = 12 + 4 = 16.$$

EXAMPLE 25. Determine, if it exists,

$$\lim_{x \to 1} \frac{x^2 + 2x - 1}{x + 1}.$$

Solution.

$$\lim_{x \to 1} (x^2 + 2x - 1) = \lim_{x \to 1} x^2 + \lim_{x \to 1} 2x + \lim_{x \to 1} (-1) = 1 + 2 - 1 = 2$$

$$\lim_{x \to 1} (x + 1) = \lim_{x \to 1} x + \lim_{x \to 1} 1 = 1 + 1 = 2$$

$$\lim_{x \to 1} \frac{x^2 + 2x - 1}{x + 1} = \frac{\lim_{x \to 1} (x^2 + 2x - 1)}{\lim_{x \to 1} (x + 1)} = \frac{2}{2} = 1$$

EXAMPLE 26. Determine, if it exists,

$$\lim_{x \to 3} \frac{x^2 - 9}{x - 3}.$$

Solution. $\lim_{x \to 3} (x^2 - 9) = 0$ and $\lim_{x \to 3} (x - 3) = 0$. Thus the limit theorem for quotients cannot be applied to determine $\lim_{x \to 3} [(x^2 - 9)/(x - 3)]$ since one of the conditions for its application is that the quotient function have a nonzero limit. However, the fact that one of the conditions for the application of the limit theorem for quotients is not satisfied does not imply that $\lim_{x \to 3} [(x^2 - 9)/(x - 3)]$ does not exist. It only means that in this situation the limit theorem for quotients cannot be used. The question can be settled by observing that

$$\frac{x^2 - 9}{x - 3} = \frac{(x - 3)(x + 3)}{(x - 3)} = x + 3$$

where $x \neq 3$. Thus

$$\lim_{x \to 3} \frac{x^2 - 9}{x - 3} = \lim_{x \to 3} (x + 3) = 6.$$

EXAMPLE 27. For $h(x) = f(x) + g(x)$, where

$$f(x) = \begin{cases} 3, & x \geq 2 \\ 1, & x < 2 \end{cases} \quad \text{and} \quad g(x) = \begin{cases} 1, & x \geq 2 \\ 3, & x < 2 \end{cases}$$

determine, if it exists, $\lim_{x \to 2} h(x)$.

Solution. If we attempt to use the limit theorem for sums, we immediately run into difficulties. $\lim_{x \to 2} f(x)$ does not exist since as x takes on values closer and closer to 2, $f(x)$ does not get closer and closer to a single value. Instead $f(x)$ clusters about two values,

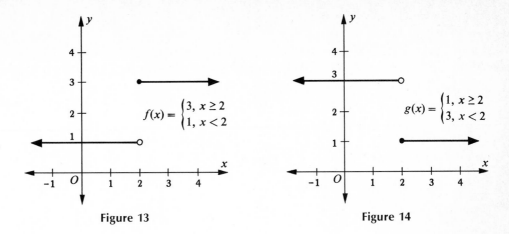

Figure 13

Figure 14

3 and 1 (see Figure 13). For the same reason, $\lim\limits_{x \to 2} g(x)$ does not exist (see Figure 14). This situation does not imply that $\lim\limits_{x \to 2} h(x)$ does not exist. All we can say is that the limit theorem for sums cannot be applied since the conditions necessary for its application are not satisfied. To determine whether or not $\lim\limits_{x \to 2} h(x)$ exists, we will examine the situation in terms of the definition of limit—the final court of appeal in matters of this kind. Let us observe that $h(x) = 4$,

$$h(x) = \begin{cases} 3 + 1 = 4, & x \geq 2 \\ 1 + 3 = 4, & x < 2 \end{cases}$$

a constant function whose graph is shown in Figure 15. By definition, $\lim\limits_{x \to 2} h(x) = 4$; as x takes on values closer and closer to 2, $h(x) = 4$ remains constant at 4.

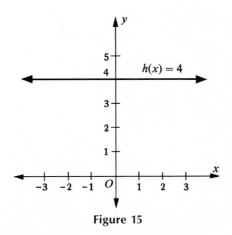

Figure 15

Exercises

Determine if the limits of the following functions exist. When a limit exists, find its value. When a limit does not exist, explain why.

58. $\lim\limits_{x \to 2} (2x^2 + 3x)$

59. $\lim\limits_{x \to 2} (3x^2 - 4x + 5)$

60. $\lim\limits_{x \to 3} (3x^3 + 2x - 4)$

61. $\lim\limits_{x \to 4} (x^3 - 2x^2 + 4x - 5)$

62. $\lim\limits_{x \to -3} \left(4x^2 - \dfrac{1}{x} + 7\right)$

63. $\lim\limits_{x \to 2} \dfrac{x^3 + 2x}{x + 1}$

64. $\lim\limits_{x \to 4} (x^3 + 2)(4x^2 + 1)$

65. $\lim\limits_{x \to 3} (2x^2 - 3x + 2)(3x^3 - 4)$

66. $\lim\limits_{x \to 3} \dfrac{3x^2 - 4x}{3x + 8}$

67. $\lim\limits_{x \to -3} \dfrac{4x^3 + x - 7}{3x^2 + 2x - 1}$

68. $\lim\limits_{x \to 3} (x^3 - 4)(2x^2 + 8)(x + 3)$

69. $\lim\limits_{x \to 2} (x^2 + 2x)(3x^2 - 1)(4x + 3)$

70. $\lim\limits_{x \to 4} \dfrac{3x^2 + 4x + 2}{2x^3 - 12}$

71. $\lim\limits_{x \to -1} \dfrac{x^2 - 2x - 1}{x - 1}$

72. $\lim\limits_{x \to 3} (x^3 + 2)(x^2 + 2x + 1)$

73. $\lim\limits_{x \to -2} \dfrac{2x^2 + x - 1}{x - 3}$

74. $\lim\limits_{x \to 50} \dfrac{f(x) - f(50)}{x - 50}$, where $f(x) = 600x - 5x^2$

75. $\lim\limits_{t \to 2} \dfrac{f(t) - f(2)}{t - 2}$, where $f(t) = 2t^2 + 1$

76. $\lim\limits_{x \to 2} \dfrac{f(x) - f(2)}{x - 2}$, where $f(x) = \dfrac{1}{2}x^2 + x + 1$

77. $\lim\limits_{p \to 50} \dfrac{f(p) - f(50)}{p - 50}$, where $f(p) = 120 - \dfrac{1}{5}p$

78. $\lim\limits_{x \to 1} \dfrac{f(x) - f(1)}{x - 1}$, where $f(x) = 2$

79. $\lim\limits_{x \to 3} \dfrac{f(x) - f(3)}{x - 3}$, where $f(x) = 100 - x^2$

80. $\lim\limits_{x \to 10} \dfrac{f(x) - f(10)}{x - 10}$, where $f(x) = \dfrac{1}{5}x^2 + 3x + 1$

81. $\lim\limits_{x \to 1} [f(x) + g(x)]$, where $f(x) = \begin{cases} 2, & x \geq 1 \\ 1, & x < 1 \end{cases}$ and $g(x) = \begin{cases} 1, & x \geq 1 \\ 2, & x < 1 \end{cases}$

82. $\lim\limits_{x \to 3} [f(x) + g(x)]$, where $f(x) = \begin{cases} 2, & x \geq 3 \\ x + 1, & x < 3 \end{cases}$ and $g(x) = \begin{cases} x^2, & x \geq 3 \\ 1 - x, & x < 3 \end{cases}$

83. Two students were asked to determine $\lim\limits_{x \to 2} [(x^4 - 16)/(x - 2)]$. Jim said, "This

problem can be solved by the limit theorem for quotients. Since

$$\lim_{x\to 2} (x^4 - 16) = 0 \qquad \text{and} \qquad \lim_{x\to 2} (x - 2) = 0$$

it follows that $\lim_{x\to 2} \dfrac{x^4 - 16}{x - 2} = \dfrac{0}{0} = 1$." Bill took issue with Jim and pointed out that "since the limit of the denominator function, $x - 2$, is 0 when x approaches 2, then $\lim_{x\to 2} \dfrac{x^4 - 16}{x - 2}$ does not exist."

a. Do you agree with Jim's analysis? Explain.
b. Do you agree with Bill's analysis? Explain.
c. If you do not agree with either analysis, how would you answer the question?

5. Continuity of a function

We have seen that $\lim_{x\to a} h(x)$ may exist even though $h(a)$ is not defined (see Example 19, Section 3) and that even when $h(a)$ is defined it may not be the same as $\lim_{x\to a} h(x)$ (see Example 20, Section 3).

If h is a function such that

1. $h(a)$ is defined
2. $\lim_{x\to a} h(x)$ exists
3. $\lim_{x\to a} h(x) = h(a)$

that is, the value of h at a and the limit of h as x approaches a are the same, then h is said to be *continuous at a*.

The function $h(x) = [(x^2 - 9)/(x - 3)]$ (discussed in Example 19, Section 3) is not continuous at 3 since condition 1 is not satisfied; $h(3)$ is not defined. See Figure 16 on the next page.
The function

$$m(x) = \begin{cases} \dfrac{x^2 - 9}{x - 3}, & x \neq 3 \\ 5, & x = 3 \end{cases}$$

(discussed in Example 20, Section 3) is not continuous at 3 since condition 3 is not satisfied; $\lim_{x\to 3} m(x) = 6$ while $m(3) = 5$, so that $\lim_{x\to 3} m(x) \neq m(3)$. See Figure 17 on the next page.
The function

$$f(x) = \begin{cases} 1, & x \geq 0 \\ -1, & x < 0 \end{cases}$$

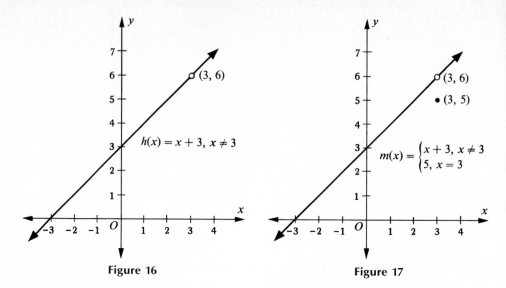

Figure 16

Figure 17

(discussed in Example 18, Section 3) is not continuous at 0 since condition 2 is not satisfied; $\lim\limits_{x \to 0} f(x)$ does not exist. See Figure 18.

The function $f(x) = 2x + 1$ (discussed in Example 17, Section 3) is continuous at 2 since $\lim\limits_{x \to 2} f(x) = 5$ and $f(2) = 5$. See Figure 19.

EXAMPLE 28. Determine if the function $f(x) = x^2$ is continuous at 2.

Solution. $\qquad f(2) = 4 \qquad$ and $\qquad \lim\limits_{x \to 2} f(x) = 4$

Thus since $\lim\limits_{x \to 2} f(x) = f(2)$, $f(x) = x^2$ is continuous at 2.

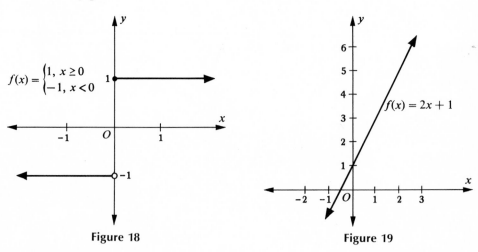

Figure 18

Figure 19

A function h is said to be *continuous over an interval I* if it is continuous at each value a in I. h is said to be *continuous over the set of real numbers* if it is continuous at each real number.

The geometric significance of continuity is that the graph of a function continuous over an interval I has no jumps, gaps, or breaks over interval I. The graph is, intuitively speaking, in one piece over I. If a function is continuous over the set of real numbers, then its graph has no jumps, gaps, or breaks.

EXAMPLE 29. Show that $f(x) = x^2$ is continuous over the set of real numbers.

Solution. Let a denote an arbitrary, but fixed, real number. Let us first observe that $f(a) = a^2$. Let us also observe that $\lim\limits_{x \to a} x^2 = a^2$. Thus $\lim\limits_{x \to a} f(x) = f(a)$, which means that $f(x) = x^2$ is continuous at a. Since a is an arbitrary real number, that is, a can be *any* real number, we can conclude that $f(x) = x^2$ is continuous at *every* real number. The geometric meaning of this is that the graph of $f(x) = x^2$ has no jumps, gaps, or breaks. See Figure 20.

EXAMPLE 30. Show that $f(x) = 1/x$ is continuous at every number in its domain of definition.

Solution. Let a denote an arbitrary, but fixed, number in the domain of definition of $f(x) = 1/x$ (thus $a \neq 0$). Since $f(a) = 1/a$ and $\lim\limits_{x \to a} (1/x) = 1/a$, it follows that $\lim\limits_{x \to a} f(x) = f(a)$. Thus $f(x) = 1/x$ is continuous at a, providing $a \neq 0$. Since a

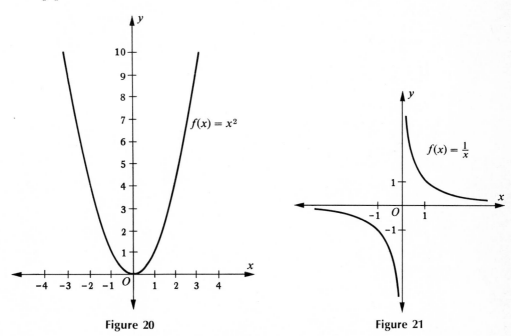

Figure 20

Figure 21

can be any value in the domain of definition of f, we conclude that f is continuous at every number in its domain of definition. Geometrically speaking, this means that the graph of $f(x) = 1/x$ has no jumps, gaps, or breaks except at 0. See Figure 21, page 25.

Exercises

84. Is $f(x) = 2x$ continuous at 2? Explain.

85. Is $f(x) = \begin{cases} x^2, & x \neq 2 \\ 1, & x = 2 \end{cases}$ continuous at 2? Explain.

86. Is $f(x) = 2x^2 + 1$ continuous at 1? Explain.

87. Is $f(x) = \dfrac{x^2 - 4}{x - 2}$ continuous at 2? Explain.

88. Is $f(x) = \begin{cases} 2, & x \geq 3 \\ x + 1, & x < 3 \end{cases}$ continuous at 3? Explain.

89. Is $f(x) = \dfrac{x^3 + 2x}{x + 1}$ continuous at 2? Explain.

90. Is $f(x) = \dfrac{2x^2 + x - 3}{x^2 + 4}$ continuous at 3? Explain.

91. Is $f(x) = (2x^2 + 1)(3x - 2)$ continuous at 2? Explain.

92. Is $f(x) = \begin{cases} x, & x \geq 0 \\ -x, & x < 0 \end{cases}$ continuous at 0? Explain.

93. Is $f(x) = \begin{cases} \dfrac{1}{x}, & x \neq 0 \\ 0, & x = 0 \end{cases}$ continuous at 0? Explain.

94. Is $f(x) = x^2 + 2x + 1$ continuous over the set of real numbers? Explain.

95. Is $f(x) = 1/x^2$ continuous at every value in its domain of definition? Explain.

6. Other limit concepts

A fundamental problem first considered in Section 3 is how does $f(x)$ behave as x takes on values closer and closer to a given value a? Here we will continue the study of this problem and consider related problems. Let us begin by examining the behavior of $f(x) = 1/x^2$ as x takes on values closer and closer to 0. From $f(x) = 1/x^2$ we obtain the following table of values. Plotting these values yields

x	$\pm\dfrac{1}{2}$	$\pm\dfrac{1}{5}$	$\pm\dfrac{1}{10}$	$\pm\dfrac{1}{100}$	± 1	± 5	± 10	± 100
$f(x)$	4	25	100	10,000	1	$\dfrac{1}{25}$	$\dfrac{1}{100}$	$\dfrac{1}{10,000}$

the graph shown in Figure 22, from which we see that as x takes on values (both positive and negative) which are closer and closer to 0, $f(x) = 1/x^2$ increases

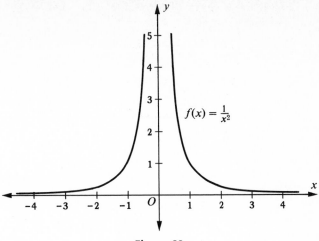

Figure 22

without bound. We can make $f(x) = 1/x^2$ as large as we please by taking x sufficiently close to 0. In this case we say that $f(x) = 1/x^2$ approaches infinity (written ∞) as x approaches 0 and express this behavior symbolically by writing

$$\operatorname*{limit}_{x \to 0} \frac{1}{x^2} = \infty \qquad \text{or} \qquad \frac{1}{x^2} \to \infty \quad \text{as} \quad x \to 0.$$

More generally, if as x takes on values closer and closer to some value a, then $f(x)$ increases without bound, we say that $f(x)$ *approaches infinity as x approaches a* and write

$$\operatorname*{limit}_{x \to a} f(x) = \infty \qquad \text{or} \qquad f(x) \to \infty \quad \text{as} \quad x \to a.$$

"$f(x)$ approaches infinity" is simply another way of saying $f(x)$ increases without bound. The symbol for infinity, ∞, does not stand for a number called "infinity." This symbol has no meaning when taken out of context. The phrase "x approaches infinity" (symbolically represented by $x \to \infty$) or "$f(x)$ approaches infinity" ($f(x) \to \infty$) is to be taken as a whole. "x approaches infinity" is a way of expressing the situation when x is increasing without bound, just as "$f(x)$ approaches infinity" expresses the situation when $f(x)$ is increasing without bound.

Let us also observe from Figure 22 that as x increases without bound (1, 5, 10, 100), $f(x) = 1/x^2$ gets closer and closer to 0 $(1, \frac{1}{25}, \frac{1}{100}, \frac{1}{10,000})$. We say that $1/x^2$ approaches 0 as x approaches infinity and write

$$\operatorname*{limit}_{x \to \infty} \frac{1}{x^2} = 0 \qquad \text{or} \qquad \frac{1}{x^2} \to 0 \quad \text{as} \quad x \to \infty.$$

More generally, if $f(x)$ gets closer and closer to a number L, or remains constant at L, as x increases without bound, then we say that the *limit of $f(x)$ is L as x approaches infinity.*

$$\operatorname*{limit}_{x \to \infty} f(x) = L \quad \text{or} \quad f(x) \to L \text{ as } x \to \infty$$

Let us further observe that as x decreases without bound $(-1, -5, -10, -100)$, $f(x) = 1/x^2$ gets closer and closer to 0 $(1, \frac{1}{25}, \frac{1}{100}, \frac{1}{10,000})$. We say that $f(x) = 1/x^2$ approaches 0 as x approaches negative infinity.

$$\operatorname*{limit}_{x \to \infty} \frac{1}{x^2} = 0 \quad \text{or} \quad \frac{1}{x^2} \to 0 \text{ as } x \to -\infty$$

More generally, if $f(x)$ gets closer and closer to a number L, or remains constant at L, as x decreases without bound, then we say that the *limit of $f(x)$ is L as x approaches negative infinity.*

$$\operatorname*{limit}_{x \to -\infty} f(x) = L \quad \text{or} \quad f(x) \to L \text{ as } x \to -\infty$$

The concepts,

$$\operatorname*{limit}_{x \to a} f(x) = -\infty \qquad \operatorname*{limit}_{x \to -\infty} f(x) = \infty$$

$$\operatorname*{limit}_{x \to \infty} f(x) = \infty \qquad \operatorname*{limit}_{x \to -\infty} f(x) = -\infty$$

$$\operatorname*{limit}_{x \to \infty} f(x) = -\infty$$

are defined in a way which is readily suggested by the previous definitions. Thus $\operatorname*{limit}_{x \to -\infty} x^2 = \infty$ since as x decreases without bound $(x \to -\infty)$, $f(x) = x^2$ increases without bound $(x^2 \to \infty)$. $\operatorname*{limit}_{x \to 0} -(1/x^2) = -\infty$ since as x approaches 0 $(x \to 0)$, $f(x) = -(1/x^2)$ decreases without bound $(-(1/x^2) \to -\infty)$.

EXAMPLE 31. Determine the behavior of

$$f(x) = \frac{x^2 + 2x + 1}{2x^2 + x - 4} \quad \text{as} \quad x \to \infty.$$

Solution. In this expression, the behavior of $f(x)$ as $x \to \infty$ is not clear since both numerator and denominator increase without bound. To overcome this difficulty, divide the numerator and denominator of $f(x)$ by x^2. Then

$$\operatorname*{limit}_{x \to \infty} \frac{x^2 + 2x + 1}{2x^2 + x - 4} = \operatorname*{limit}_{x \to \infty} \frac{1 + \dfrac{2}{x} + \dfrac{1}{x^2}}{2 + \dfrac{1}{x} - \dfrac{4}{x^2}} = \frac{1}{2}$$

since $\dfrac{2}{x} \to 0$, $\dfrac{1}{x^2} \to 0$, $\dfrac{1}{x} \to 0$, and $-\dfrac{4}{x^2} \to 0$ as $x \to \infty$.

Exercises

Determine the limit behavior of each of the following functions.

96. $f(x) = 1 + \dfrac{1}{x}$ as $x \to \infty$

97. $f(x) = 3 - \dfrac{2}{x^2}$ as $x \to 0$

98. $f(x) = \dfrac{1}{(x-1)^2}$ as $x \to 1$

99. $f(x) = 2 + \dfrac{2}{x^3}$ as $x \to -\infty$

100. $f(x) = 5 + \dfrac{2}{(x+2)^2}$ as $x \to -2$

101. $f(x) = \dfrac{1}{x}$ as $x \to 0$

102. $f(x) = 4 + \dfrac{1}{x^4}$ as $x \to 0$

103. $f(x) = \dfrac{x^3 + 2x + 1}{3x^2 + x + 4}$ as $x \to \infty$

104. $f(x) = \dfrac{3x^2 + 1}{4x^2 + 2}$ as $x \to -\infty$

105. $f(x) = 4 + \dfrac{8}{x^2}$ as $x \to 0$

106. $f(x) = \dfrac{1}{x+1}$ as $x \to -1$

107. $f(x) = 1 + \dfrac{3}{x^5}$ as $x \to 0$

108. $f(x) = 3 - \dfrac{4}{x}$ as $x \to \infty$

109. $f(x) = \dfrac{4x^2 - 3x + 4}{7x^2 + 4x - 8}$ as $x \to \infty$

110. $f(x) = \dfrac{3x^2 + 2x + 6}{2x^2 - x + 4}$ as $x \to \infty$

111. Determine

$$\lim_{L \to \infty} \left[2K - \frac{K(KL + 1)}{L + 1} \right]$$

where K is a constant. This problem arises in connection with Robert Fein-schreiber's discussion "Accelerated Depreciation: A Proposed New Method," *Journal of Accounting Research* (Spring 1969), pp. 17–21.

Sec 6   Other limit concepts

7. A limit problem in the world of finance

In June, 1972, a number of banks in the New York City region announced that as of July 1, 1972, interest on savings accounts would be compounded continuously. Continuous compounding of interest is based on a limit concept, and in this section we will examine its nature.

Let us recall that in the world of finance *interest* is money charged for the use of borrowed money. It is an amount which is stated in terms of some monetary unit (dollars, cents, pounds). Interest should not be confused with interest rate, which is a number. *Interest rate* is the ratio of the interest charged during the interest period to the amount of money owed at the beginning of the interest period. It is usually stated as a percentage. The length of the interest period must be stated or understood. Often in colloquial language the term "interest rate" is abbreviated to "interest."

Simple interest is the interest produced in the first interest period multiplied by the number of interest periods. Thus if the amount borrowed is $100 at 8% per year for two years, the simple interest is

$$\$100(0.08)2 = \$16.$$

Compound interest is interest which is computed periodically on the total amount owed at the beginning of each interest period. Thus if $100 is borrowed for two years at the rate of 8% per year, the compound interest is

$$100(0.08) + 108(0.08) = \$16.64.$$

During the second year the amount owed is $108. The $8 in interest owed for the first year is added to the amount owed for the second year.

In financial transactions an interest rate quoted as *r compounded m times a year* is understood to mean a rate of $i = r/m$ for each of the m interest periods the year is divided into. Rate r is called a *nominal rate*. Thus a nominal rate of 6% compounded twice a year (semiannually) is by definition a rate of 3% for each six-month period; a nominal rate of 12% compounded twelve times a year is by definition a rate of 1% for each one-month period.

Suppose we wish to know the rate v compounded annually which yields the same interest at the end of the year as investment at the nominal rate of 12% compounded twelve times a year. Such a rate v is called an effective rate. More generally, for a given nominal rate r compounded m times a year, the corresponding *effective rate* v is that rate which if compounded annually would yield the same interest. In the mathematics of finance (see Appendix 2) it is shown that a nominal rate r compounded m times a year is equivalent to an effective rate v given by

$$v = \left(1 + \frac{r}{m}\right)^m - 1.$$

Thus the effective rate v compounded annually which yields the same interest at the end of the year as investment at the rate of 12% compounded twelve times a year is

$$v = \left(1 + \frac{0.12}{12}\right)^{12} - 1 = 1.127 - 1 = 12.7\%.$$

Interest rates are considered with respect to a time period. But what constitutes a time period? Must it be a year, or a month, or a week? Can it be a second? Can we regard interest accumulation as a continuous process going on every instant? Such questions lead us to study the behavior of various interest relationships as the number of interest periods m is allowed to get larger and larger without bound ($m \to \infty$).

First let us examine the behavior of

$$v = \left(1 + \frac{r}{m}\right)^m - 1$$

as $m \to \infty$. Basic to this examination is the behavior of

$$\left(1 + \frac{1}{m}\right)^m$$

as $m \to \infty$. The values of this function (correct to three decimal places) for certain values of m are shown in the table. Thus it would seem that $\left(1 + \frac{r}{m}\right)^m$ approaches

m	1	4	10	100	1000	10,000
$\left(1 + \frac{1}{m}\right)^m$	2	2.441	2.594	2.705	2.717	2.718

some value in the neighborhood of 2.718 as $m \to \infty$. This is indeed the case; the limit of $\left(1 + \frac{1}{m}\right)^m$ as $m \to \infty$ is 2.71828, correct to five decimal places, which is denoted by the letter e. Therefore we have

$$\lim_{m \to \infty} \left(1 + \frac{1}{m}\right)^m = e \tag{1}$$

where $e = 2.71828$, correct to five decimal places.

Returning to the more general case of $v + 1 = \left(1 + \frac{r}{m}\right)^m$, let $n = \frac{m}{r}$, so that $\frac{r}{m} = \frac{1}{n}$ and $m = nr$. Then

$$\left(1 + \frac{r}{m}\right)^m = \left(1 + \frac{1}{n}\right)^{nr} = \left[\left(1 + \frac{1}{n}\right)^n\right]^r.$$

As $m \to \infty$, $\frac{m}{r} \to \infty$, so $n \to \infty$, and from (1),

$$\left[\left(1 + \frac{1}{n}\right)^n\right]^r \to e^r.$$

Since $v + 1 = \left(1 + \dfrac{r}{m}\right)^m$, then as $m \to \infty$, $(v + 1) \to e^r$. Thus

$$\lim_{m \to \infty} v = \lim_{m \to \infty} \left[\left(1 + \frac{r}{m}\right)^m - 1\right]$$

$$= \lim_{m \to \infty} \left(1 + \frac{r}{m}\right)^m + \lim_{m \to \infty} (-1)$$

$$= e^r - 1.$$

The value $e^r - 1$ is defined as the *effective interest rate corresponding to a nominal rate of r compounded continuously.* A rate of r compounded continuously will yield the same interest at the end of the year as the effective interest rate $(e^r - 1)$ compounded annually.* Thus, for example, a rate of 5% compounded continuously will yield the same interest at the end of the year as the effective rate $(e^{.05} - 1) = (1.0513 - 1) = 5.13\%$ compounded annually.

In the mathematics of finance it is shown that an amount A invested at the nominal rate r compounded m times a year accumulates to the asset value

$$A\left(1 + \frac{r}{m}\right)^{mx} = A\left[\left(1 + \frac{r}{m}\right)^m\right]^x$$

after x years (see Appendix 2). Under conditions of continuous compounding of interest, where m is allowed to increase without bound $(m \to \infty)$, we have

$$A\left[\left(1 + \frac{r}{m}\right)^m\right]^x \to A\, e^{rx}$$

$$A\left(1 + \frac{r}{m}\right)^{mx} = A\left[\left(1 + \frac{r}{m}\right)^m\right]^x$$

$$A\left(1 + \frac{r}{m}\right)^{mx} \to A\, e^{rx}.$$

Thus the function

$$a(x) = A\, e^{rx}\,.$$

expresses the asset value that an initial amount of money A will grow to be in x years if invested at rate r compounded continuously. For example, if an initial amount of $1000 is invested at a rate of 5% compounded continuously, then in x years it will grow to an asset value of

$$a(x) = 1000\, e^{0.05x} \text{ dollars.}$$

---

*Institutions which operate on the basis of a modified year adjust this result accordingly. Banks, for example, operate on the basis of a 360 day year and use $e^{(365/360)r} - 1$ to determine the effective interest rate corresponding to a nominal rate of r compounded continuously. Thus the effective rates obtained by banks are higher than effective rates calculated from $e^r - 1$. A rate of 5% compounded continuously, for example, is equivalent to an effective rate of 5.13% compounded annually in terms of the standard year and 5.20% compounded annually in terms of the bank year.

Thus in one year $1000 would grow to

$$a(1) = \$1000\ e^{0.05(1)} = \$1000(1.0513) = \$1051.30.$$

The exponential function

$$a(x) = A\ e^{rx}$$

is encountered not only in the study of the growth of principal at a nominal rate r under conditions of continuous compounding of interest but also in the study of growth situations in general (population, wealth, capital).

It is also shown in the mathematics of finance (see Appendix 2) that the principal that must be invested at a nominal rate r compounded m times a year in order to grow to a given amount A in x years is

$$A\left(1 + \frac{r}{m}\right)^{-mx} = A\left[\left(1 + \frac{r}{m}\right)^{m}\right]^{-x}.$$

Under conditions of continuous compounding of interest, where $m \to \infty$,

$$A\left[\left(1 + \frac{r}{m}\right)^{m}\right]^{-x} \to A\ e^{-rx}.$$

Therefore the function

$$b(x) = A\ e^{-rx}$$

expresses the amount that must initially be invested at rate r compounded continuously if at the end of x years amount A is to be available. Thus if $2000 is to be available in x years, the amount that must be initially invested at 5% compounded continuously is

$$b(x) = 2000\ e^{-0.05x} \text{ dollars.}$$

Specifically, if $2000 is to be available in 2 years, the amount that must be initially invested at 5% compounded continuously is

$$b(2) = 2000\ e^{-0.05(2)} = 2000(0.9048) = \$1809.60.$$

Exercises

112. Determine the effective interest rate under conditions of continuous compounding of interest for the following nominal rates r (values of e^x and e^{-x} are given at the end of this series of exercises).
    a. $r = 6\%$ b. $r = 7\%$ c. $r = 8\%$ d. $r = 9\%$
113. $5000 is invested at a rate of 7% compounded continuously. Determine the function which describes the amount this sum will grow to in x years. Determine the asset value of $5000 in one year; two years; three years; four years.
114. $10,000 is invested at a rate of 9% compounded continuously. Determine the function which describes the amount this sum will grow to in x years. Determine the asset value of $10,000 in one year; two years; three years; four years.

115. Determine the function which describes the amount which must be initially invested at a rate of 8% compounded continuously if at the end of x years a certain company is to have $25,000 available for the purchase of new equipment. Determine the amount which must be invested under the stated conditions if $25,000 is to be available at the end of two years; three years; four years.

116. Determine the function which describes the amount which must be initially invested at a rate of 6% compounded continuously if at the end of x years a certain company is to have available $15,000 to pay off a debt. Determine the amount which must be invested under the stated conditions if $15,000 is to be available at the end of two years; three years; four years.

117. Determine $\lim\limits_{L \to \infty} \left[1 - \left(1 - \dfrac{2}{L} \right)^{KL} \right]$, where K is constant. This problem arises in connection with Robert Feinschreiber's discussion "Accelerated Depreciation: A Proposed New Method," *Journal of Accounting Research* (Spring 1969), pp. 17–21.

x	e^x	x	e^x	x	e^{-x}
0.06	1.0618	0.18	1.1972	0.12	0.8869
0.07	1.0725	0.21	1.2337	0.16	0.8521
0.08	1.0833	0.27	1.3100	0.18	0.8353
0.09	1.0942	0.28	1.3231	0.24	0.7866
0.14	1.1503	0.36	1.4333	0.32	0.7261

2

Basic differential calculus

8. Limit concepts with a common structure

In this section several limit concepts will be presented, some of which may already be familiar. As we will see, all the concepts have a common structural feature. By examining the various situations, we will be able to extract the common features and build a mathematical model that will fit all similar situations.

A problem in the study of motion

Let us consider a stone which is dropped from a building to the street below. If all forces acting on the stone outside of gravity are neglected, then the relationship between the distance d (in feet) that the stone falls and the time t (in seconds) that it takes to fall d feet is given by the function

$$d = 16t^2.$$

What is the instantaneous velocity of the stone at time $t = 1$ second?

The real problem here is to define what is meant by instantaneous velocity. In formulating a definition of this concept, we will take as our starting point the concept of average velocity. Let us recall that the average velocity of an object in motion is the ratio

$$\frac{\text{distance moved by the object}}{\text{time taken to move the distance}}.$$

Consider the time interval from t seconds to 1 second. The distance the stone falls in this time interval is the distance it falls in 1 second minus the distance it falls in t seconds, that is,

$$16(1)^2 - 16t^2 = 16 - 16t^2.$$

The time interval is $1 - t$. Thus the average velocity of the stone over the time interval from t seconds to 1 second is given by the function

$$a(t) = \frac{16 - 16t^2}{1 - t}.$$

By factoring and using basic algebra we simplify $a(t)$ to

$$a(t) = \frac{16 - 16t^2}{1 - t} = \frac{16(1 - t)(1 + t)}{(1 - t)} = 16 + 16t$$

where $t \neq 1$. It's obvious, but still worthy of mention, that for the concept of instantaneous velocity at time $t = 1$ to make sense, it must be a unique value. But as t takes on different values near 1, $a(t)$ takes on different values. Given the circumstances, it seems reasonable to examine the behavior of $a(t)$ as t takes on values closer and closer to 1. Let us observe that $a(t) = 16 + 16t$ gets closer and closer to 32, that is,

$$\lim_{t \to 1} a(t) = \lim_{t \to 1} (16 + 16t) = 32.$$

In physics it has been found to be both natural and fruitful to define the instantaneous velocity of the object at $t = 1$ second to be the limit of the average velocity function $a(t)$ as t approaches 1. Thus the instantaneous velocity of the stone at time $t = 1$ second is 32.

More generally, consider an object whose motion is described by a time-distance function $d = f(t)$. The average velocity of this object in the time interval from t seconds to a seconds is

$$a(t) = \frac{f(a) - f(t)}{a - t}.$$

Multiplying numerator and denominator of $a(t)$ by -1 yields

$$a(t) = \frac{f(t) - f(a)}{t - a}.$$

The *instantaneous velocity of this object at time $t = a$ seconds* is defined as

$$\lim_{t \to a} a(t) = \lim_{t \to a} \frac{f(t) - f(a)}{t - a}$$

provided that this limit exists and is a real number.

The concept of instantaneous velocity gives us a motion characteristic of an object at a certain instant of time as opposed to over an interval of time. Within the world of physics it helps us understand and predict the behavior of bodies in motion. In this respect it has been decidedly successful.

From the definition of $\lim_{x \to a} f(x) = L$ we know that when x is close to a, $f(x)$ is close to L. That is, L can be taken as an estimate of $f(x)$ for x close to a. Applying this observation to the definition of instantaneous velocity at time a,

$$\lim_{t \to a} \frac{f(t) - f(a)}{t - a} = K$$

we obtain

$$\frac{f(t) - f(a)}{t - a} \simeq K$$

for t close to a. That is, the instantaneous velocity K at time $t = a$ can be used as an estimate of the average velocity from time t to time a, provided that this time interval is small.

EXAMPLE 1. Find the instantaneous velocity at time $t = 2$ seconds of the object whose motion is described by the time-distance function $d = f(t) = 2t^2 + 1$.

Solution. The average velocity of the object in the interval from t seconds to 2 seconds is

$$\frac{f(t) - f(2)}{t - 2} = \frac{(2t^2 + 1) - 9}{t - 2} = \frac{2t^2 - 8}{t - 2} = \frac{2(t - 2)(t + 2)}{(t - 2)} = 2(t + 2)$$

where $t \neq 2$. Thus the instantaneous velocity of the object at time $t = 2$ seconds is

$$\lim_{t \to 2} \frac{f(t) - f(2)}{t - 2} = \lim_{t \to 2} 2(t + 2) = 8.$$

This numerical value is a motion characteristic of the object with respect to time $t = 2$ seconds. For small time intervals of $t - 2$ (or $2 - t$), 8 feet per second is the approximate value of the average velocity of the object.

A problem in geometry*

A classical problem which played an important role in the development of the differential calculus was that of finding the tangent line to a curve at a given point.

Let us begin our discussion of this problem by examining the special case of the curve C which is the graph of $f(x) = x^2$ (see Figure 1). Consider two specific questions. How should the tangent line to C at the origin $(0, 0)$ be defined? How should the tangent line to C at the point $(2, 4)$ be defined? The nature of the situation makes it natural to define the x-axis as the tangent line to C at $(0, 0)$. Unfortunately this solution does not provide us with an approach to the general definition of the tangent line. It provides us with no insight, for example, into how to define the

*This discussion presupposes knowledge of the straight line, which is reviewed in Appendix 1.

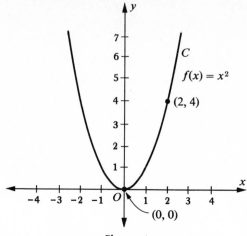

Figure 1

tangent line to C at $(2, 4)$. What we will do is use the problem of defining the tangent line to C at $(0, 0)$ as a starting point for the development of a definition for the tangent line which can be used in general.

Let $Q(x, x^2)$ denote a point on C which is near the origin O, and let OQ denote the line through O and Q (see Figure 2). Let us observe that as we vary the position of Q on C so that Q approaches origin O, the line OQ approaches the x-axis—our candidate for the tangent line to C at O. How does the slope of OQ behave?

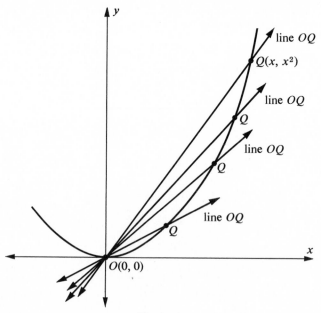

Figure 2

The slope of OQ is described by the function

$$m(x) = \frac{x^2 - 0}{x - 0} = x$$

where $x \neq 0$. As Q approaches O on C, x approaches zero. As x approaches zero, the slope function $m(x) = x$ approaches zero, that is,

$$\lim_{x \to 0} m(x) = 0.$$

But let us note that zero is the slope of the x-axis, our candidate for the tangent line to C at the origin. Thus if we define the tangent line to C at the origin to be that line through the origin whose slope is

$$\lim_{x \to 0} m(x) = \lim_{x \to 0} x = 0$$

then we obtain the x-axis as the tangent line.

This approach can be carried over to other situations. Consider $P(2, 4)$ on C and let $Q(x, x^2)$ denote a point on C which is near P; let PQ denote the line through P and Q (see Figure 3). The slope of PQ is described by the slope function

$$m(x) = \frac{x^2 - 4}{x - 2} = \frac{(x - 2)(x + 2)}{(x - 2)} = x + 2$$

where $x \neq 2$. As $Q(x, x^2)$ approaches $P(2, 4)$ on C, x approaches 2; as x approaches

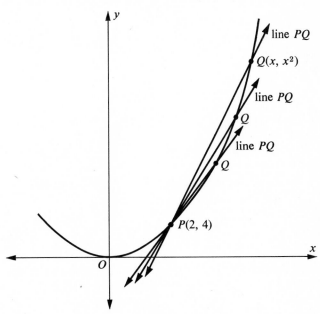

Figure 3

2, the slope function $m(x) = x + 2$ approaches 4; that is,

$$\underset{x \to 2}{\text{limit}}\, m(x) = \underset{x \to 2}{\text{limit}}\, (x + 2) = 4.$$

We define as the tangent line to C at $P(2, 4)$ that line through $P(2, 4)$ whose slope is 4. An equation of this line is

$$y - 4 = 4(x - 2)$$

which simplifies to

$$y = 4x - 4.$$

More generally, consider a curve C which is the graph of a function $y = f(x)$ and a point $P(a, f(a))$ on C. Let $Q(x, f(x))$ denote a point on C which is near P (see Figure 4), and let PQ denote the line through P and Q. The slope of PQ is described by the slope function

$$m(x) = \frac{f(x) - f(a)}{x - a}.$$

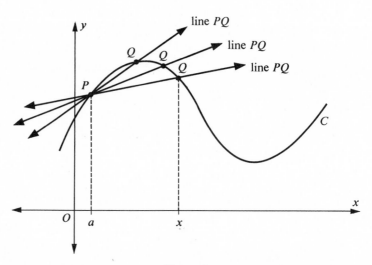

Figure 4

As Q approaches P on C, x approaches a. If as x approaches a, $m(x)$ approaches a number M, that is,

$$\underset{x \to a}{\text{limit}}\, m(x) = \underset{x \to a}{\text{limit}}\, \frac{f(x) - f(a)}{x - a} = M$$

then we define the *tangent line to C at $P(a, f(a))$ to be the line through P whose slope is M.*

EXAMPLE 2.   Find an equation for the tangent line to the graph of $f(x) = 2x^2 + 1$ at the point $P(2, 9)$.

Solution.   By definition, the tangent line to the given curve at $P(2, 9)$ is the line through $P(2, 9)$ whose slope is

$$\operatorname*{limit}_{x \to 2} \frac{f(x) - f(2)}{x - 2}$$

provided that this limit exists and is a real number.

$$\frac{f(x) - f(2)}{x - 2} = \frac{(2x^2 + 1) - 9}{x - 2} = \frac{2x^2 - 8}{x - 2} = \frac{2(x - 2)(x + 2)}{(x - 2)} = 2(x + 2)$$

Thus

$$\operatorname*{limit}_{x \to 2} \frac{f(x) - f(2)}{x - 2} = \operatorname*{limit}_{x \to 2} 2(x + 2) = 8.$$

Therefore the tangent line to the graph of $f(x) = 2x^2 + 1$ at $P(2, 9)$ has slope 8. (See Figure 5.) An equation of this tangent line is

$$y - 9 = 8(x - 2)$$

which simplifies to

$$y = 8x - 7.$$

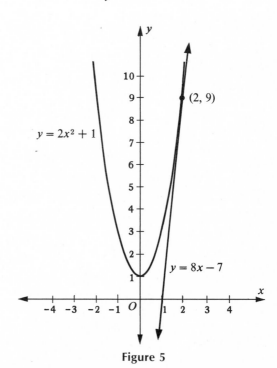

Figure 5

A situation in economics

Consider a monopolist who is supplying a certain commodity to a market and suppose that the total revenue derived by the monopolist in supplying x units to the market per unit time is given by the revenue function

$$R = f(x).$$

If the monopolist supplies a units per unit time to the market, his revenue is $f(a)$ monetary units. If output is increased from a units to x units, the change in revenue is given by the difference

$$f(x) - f(a).$$

In economics one approach to the study of revenue that has proved fruitful is based on the study of the average change in revenue for a small change in output, that is,

$$\frac{f(x) - f(a)}{x - a}$$

for small changes in output from a units to x units. The foundation for this study is provided by the concept of *marginal revenue for an output of a units,* which is defined as

$$\lim_{x \to a} \frac{f(x) - f(a)}{x - a} = L$$

provided that this limit exists and L is a real number.

The concept of marginal revenue gives us a production characteristic for an economic enterprise with respect to a certain output level. In the world of business and economics it helps us understand and predict the behavior of economic structures. In this respect it has proved fruitful. The significance of marginal revenue in connection with the profit-maximization behavior of a firm is examined in Section 21.

From the definition of limit, it follows that

$$\lim_{x \to a} \frac{f(x) - f(a)}{x - a} = L$$

means that when x is close to a,

$$\frac{f(x) - f(a)}{x - a} \simeq L$$

or equivalently

$$f(x) - f(a) \simeq L(x - a)$$

for x close to a. That is, the increase in revenue brought about by a small increase in output from a units to x units can be approximated by the product of L, the marginal revenue for output a, times the increase in output $x - a$.

To illustrate, suppose that the total dollar revenue derived by a sugar refinery in supplying x tons of sugar to the market per week is

$$R = f(x) = 600x - 5x^2.$$

Then the marginal revenue for an output of 50 tons a week is

$$\underset{x \to 50}{\text{limit}} \frac{f(x) - f(50)}{x - 50}.$$

$$\frac{f(x) - f(50)}{x - 50} = \frac{600x - 5x^2 - 17{,}500}{x - 50} = \frac{-5(x^2 - 120x + 3500)}{x - 50}$$

$$= \frac{-5(x - 50)(x - 70)}{x - 50} = -5(x - 70)$$

Thus the marginal revenue for an output of 50 tons of sugar a week is

$$\underset{x \to 50}{\text{limit}} \frac{f(x) - f(50)}{x - 50} = \underset{x \to 50}{\text{limit}} -5(x - 70) = 100.$$

We say that total revenue is increasing at the instantaneous rate of $100 per ton of sugar when 50 tons of sugar a week are being produced.

Another situation in economics

Consider a firm which is producing a certain commodity and suppose that the total cost of producing x units per unit time is given by the cost function

$$c = f(x).$$

If the firm produces a units per unit time, its cost is $f(a)$ monetary units. If output is increased from a units to x units, the increase in cost is given by the difference

$$f(x) - f(a).$$

The ratio

$$\frac{f(x) - f(a)}{x - a}$$

expresses the average cost of the additional output. In economics it has proved fruitful to study this ratio when the increase in output from a units to x units is small. The basis for this approach is the concept of *marginal cost for an output of a units,* defined as

$$\underset{x \to a}{\text{limit}} \frac{f(x) - f(a)}{x - a} = K$$

provided that this limit exists and K is a real number.

The concept of marginal cost gives us another production characteristic for an economic enterprise with respect to a certain output level. In the world of business

and economics it helps us understand and predict the behavior of economic structures. The significance of marginal cost in connection with the profit-maximization
behavior of a firm is examined in Section 21.

From the definition of limit, it follows that

$$\lim_{x \to a} \frac{f(x) - f(a)}{x - a} = K$$

means that when x is close to a,

$$\frac{f(x) - f(a)}{x - a} \simeq K$$

or equivalently

$$f(x) - f(a) \simeq K(x - a)$$

for x close to a. That is, the increase in cost brought about by a small increase
in output from a units to x units can be approximated by the product of K, the
marginal cost for output a, times the increase in output $x - a$.

To illustrate, suppose that the total cost of producing x tons of sugar a week is
given by the cost function

$$c = f(x) = \frac{1}{2}x^2 + x + 1.$$

Then the marginal cost for an output of 2 tons of sugar a week is

$$\lim_{x \to 2} \frac{f(x) - f(2)}{x - 2}.$$

$$\frac{f(x) - f(2)}{x - 2} = \frac{(\frac{1}{2}x^2 + x + 1) - 5}{x - 2} = \frac{\frac{1}{2}x^2 + x - 4}{x - 2}$$

$$= \frac{(x - 2)(\frac{1}{2}x + 2)}{x - 2} = \frac{1}{2}x + 2$$

Thus the marginal cost of an output of 2 tons of sugar a week is

$$\lim_{x \to 2} \frac{f(x) - f(2)}{x - 2} = \lim_{x \to 2} (\tfrac{1}{2}x + 2) = 3.$$

We say that total cost is increasing at an instantaneous rate of \$3 per ton of sugar
when 2 tons of sugar a week are being produced.

9. The concept of derivative

In connection with our discussion of motion we saw that the instantaneous velocity
at time $t = a$ seconds of an object whose motion is described by the time-distance
function $d = f(t)$ is defined as

$$\operatorname*{limit}_{t \to a} \frac{f(t) - f(a)}{t - a}.$$

In considering a problem in geometry we observed that the tangent line to the graph of a function $y = f(x)$ at the point $P(a, f(a))$ is defined as the line through P whose slope is

$$\operatorname*{limit}_{x \to a} \frac{f(x) - f(a)}{x - a}.$$

In economics, if $R = f(x)$ is the revenue function for a monopolist supplying x units of some commodity to a market, then the marginal revenue for an output of a units is defined as

$$\operatorname*{limit}_{x \to a} \frac{f(x) - f(a)}{x - a}.$$

If $c = f(x)$ is the total cost function for a firm producing x units of some commodity, then the marginal cost for an output of a units is defined as

$$\operatorname*{limit}_{x \to a} \frac{f(x) - f(a)}{x - a}.$$

These are only four of many different situations which present a common structural feature. Let us identify this structural feature and study it in its own right without reference to the various backgrounds in which it is to be found. We can later return to these different backgrounds and apply to them whatever results we have obtained.

In each of the four situations, we start with a function, which we will here denote by $y = f(x)$, and a value a at which this function is defined. This function is, of course, interpreted differently in the different situations. In one instance it is thought of as a time-distance function, while in another situation it is viewed as a total cost function. No matter, the feature common to all is that we have a function $y = f(x)$ and a value a at which it is defined.

We next construct a new function

$$g(x) = \frac{f(x) - f(a)}{x - a}.$$

This function is also interpreted differently in the different situations.

Our final step in all of these situations is to form

$$\operatorname*{limit}_{x \to a} \frac{f(x) - f(a)}{x - a}.$$

This extraction of the structural feature common to all the situations we have examined leads us to a concept called the derivative of a function.

If f is a function defined at a number a, then the function f', whose value at a is

$$f'(a) = \operatorname*{limit}_{x \to a} \frac{f(x) - f(a)}{x - a}$$

whenever this limit exists and is a real number, is called the *derivative of f*. The value $f'(a)$ is called the *derivative of f at a*. If $f'(a)$ exists, then f is said to be *differentiable at a*.

When the notation $y = f(x)$ is used for the function f, with y denoting the dependent variable and x the independent variable, then $f'(a)$ is often referred to as the *instantaneous rate of change* (or *rate of change*) of y with respect to x at a. This terminology usually appears in connection with applications of the derivative.

EXAMPLE 3. Find $f'(3)$ for $f(x) = 100 - x^2$ and interpret the result.

Solution. By definition, $f'(3) = \underset{x \to 3}{\text{limit}} \dfrac{f(x) - f(3)}{x - 3}$.

$$\frac{f(x) - f(3)}{x - 3} = \frac{(100 - x^2) - 91}{x - 3} = \frac{-x^2 + 9}{x - 3} = \frac{-1(x - 3)(x + 3)}{(x - 3)} = -x - 3$$

Thus

$$f'(3) = \underset{x \to 3}{\text{limit}} \frac{f(x) - f(3)}{x - 3} = \underset{x \to 3}{\text{limit}} (-x - 3) = -6.$$

If $f(x) = 100 - x^2$ is interpreted as expressing a time-distance relationship for an object in motion, then $f'(3) = -6$ represents the instantaneous velocity of the object at time 3. Geometrically speaking, $f'(3) = -6$ is the slope of the tangent line to the graph of $f(x) = 100 - x^2$ at the point $P(3, f(3))$, or $P(3, 91)$. If $f(x) = 100 - x^2$ is interpreted as expressing the revenue function for a monopolist supplying x units of some commodity to a market, then $f'(3) = -6$ represents the marginal revenue for an output of 3 units. If $f(x) = 100 - x^2$ is interpreted as the total cost function for the production of x units of some commodity, then $f'(3) = -6$ is the marginal cost for an output of 3 units.

EXAMPLE 4. For $f(x) = \frac{1}{5}x^2 + 3x + 1$, determine $f'(10)$.

Solution. By definition, $f'(10) = \underset{x \to 10}{\text{limit}} \dfrac{f(x) - f(10)}{x - 10}$.

$$\frac{f(x) - f(10)}{x - 10} = \frac{(\frac{1}{5}x^2 + 3x + 1) - 51}{x - 10} = \frac{\frac{1}{5}x^2 + 3x - 50}{x - 10}$$

$$= \frac{x^2 + 15x - 250}{5(x - 10)} = \frac{(x - 10)(x + 25)}{5(x - 10)} = \frac{1}{5}(x + 25)$$

Thus

$$f'(10) = \underset{x \to 10}{\text{limit}} \frac{f(x) - f(10)}{x - 10} = \underset{x \to 10}{\text{limit}} \frac{1}{5}(x + 25) = 7.$$

EXAMPLE 5.   Determine f' for the constant function $f(x) = 2$.

Solution.   Let a be a value in the domain of definition of f'. Then for $x \neq a$,

$$\frac{f(x) - f(a)}{x - a} = \frac{2 - 2}{x - a} = 0.$$

Thus

$$f'(a) = \lim_{x \to a} \frac{f(x) - f(a)}{x - a} = \lim_{x \to a} 0 = 0.$$

In stating our final conclusion, we can use any symbol we wish for the independent variable of f'. When x is the independent variable of f, it is customary to also use x for the independent variable of f'. That is, for $f(x) = 2$, we would use the notation $f'(x) = 0$ for the derivative of $f(x)$.

EXAMPLE 6.   Determine f' for $f(x) = x$.

Solution.   Let a be a value in the domain of definition of f'. Then for $x \neq a$,

$$\frac{f(x) - f(a)}{x - a} = \frac{x - a}{x - a} = 1.$$

Thus

$$f'(a) = \lim_{x \to a} \frac{f(x) - f(a)}{x - a} = \lim_{x \to a} 1 = 1.$$

Therefore for $f(x) = x, \; f'(x) = 1$.

EXAMPLE 7.   Determine f' for $f(x) = x^2$.

Solution.   Let a be a value in the domain of definition of f'. Then for $x \neq a$,

$$\frac{f(x) - f(a)}{x - a} = \frac{x^2 - a^2}{x - a} = \frac{(x - a)(x + a)}{x - a} = x + a.$$

Thus

$$f'(a) = \lim_{x \to a} \frac{f(x) - f(a)}{x - a} = \lim_{x \to a} (x + a) = 2a.$$

Therefore for $f(x) = x^2, f'(x) = 2x$.

EXAMPLE 8.   Determine f' for $f(x) = x^3$.

Solution.   Let a be a value in the domain of definition of f'. Then for $x \neq a$,

$$\frac{f(x) - f(a)}{x - a} = \frac{x^3 - a^3}{x - a}$$

$$= \frac{(x - a)(x^2 + ax + a^2)}{x - a}$$

$$= x^2 + ax + a^2.$$

Thus

$$f'(a) = \operatorname*{limit}_{x \to a} \frac{f(x) - f(a)}{x - a} = \operatorname*{limit}_{x \to a} (x^2 + ax + a^2)$$

$$= \operatorname*{limit}_{x \to a} x^2 + \operatorname*{limit}_{x \to a} ax + \operatorname*{limit}_{x \to a} a^2$$

$$= a^2 + a^2 + a^2 = 3a^2.$$

Therefore for $f(x) = x^3$, $f'(x) = 3x^2$.

EXAMPLE 9.   Determine f' for $f(x) = x^4$.

Solution.   Let a be a value in the domain of definition of f'. Then for $x \neq a$,

$$\frac{f(x) - f(a)}{x - a} = \frac{x^4 - a^4}{x - a} = \frac{(x^2 - a^2)(x^2 + a^2)}{x - a}$$

$$= \frac{(x - a)(x + a)(x^2 + a^2)}{x - a} = (x + a)(x^2 + a^2).$$

Thus

$$f'(a) = \operatorname*{limit}_{x \to a} \frac{f(x) - f(a)}{x - a} = \operatorname*{limit}_{x \to a} (x + a)(x^2 + a^2)$$

$$= \operatorname*{limit}_{x \to a} (x + a) \cdot \operatorname*{limit}_{x \to a} (x^2 + a^2)$$

$$= 2a \cdot 2a^2 = 4a^3.$$

Therefore for $f(x) = x^4$, $f'(x) = 4x^3$.

EXAMPLE 10.   Determine, if it exists, $f'(0)$ for $f(x) = \begin{cases} x, & x \geq 0 \\ -x, & x < 0 \end{cases}$.

Solution.   By definition,

$$f'(0) = \operatorname*{limit}_{x \to 0} \frac{f(x) - f(0)}{x - 0}$$

provided that this limit exists.

$$g(x) = \frac{f(x) - f(0)}{x - 0} = \frac{f(x) - 0}{x} = \frac{f(x)}{x}, \quad x \neq 0$$

$$g(x) = \frac{f(x)}{x} = \begin{cases} \dfrac{x}{x} = 1, & x > 0 \\ \dfrac{-x}{x} = -1, & x < 0 \end{cases}$$

$$g(x) = \begin{cases} 1, & x > 0 \\ -1, & x < 0 \end{cases}$$

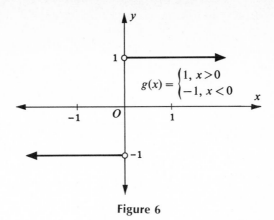

Figure 6

The graph of $g(x)$ is shown in Figure 6.

$$f'(0) = \text{limit}_{x \to 0} \frac{f(x) - f(0)}{x - 0} = \text{limit } g(x)$$

which does not exist. As x takes on values closer and closer to 0, $g(x)$ does not get closer and closer to a number. Instead, $g(x)$ clusters about two values, 1 and -1 (see Figure 6). Thus $f'(0)$ does not exist.

In our discussion of the derivative concept, we used f' and $f'(x)$ to denote the derivative function of a function f. $f'(x)$ is also used to denote the value of the derivative function f' at x, just as $f'(a)$ is used to denote the value of f' at a. Other symbols for the derivative of f are Df, $Df(x)$, $\dfrac{df}{dx}$, and $\dfrac{df(x)}{dx}$. When y is used for the dependent variable of function f, that is, $y = f(x)$, then y', $y'(x)$, and $\dfrac{dy}{dx}$ are also used to denote the derivative of f.

Exercises

Use the definition of derivative to determine the derivatives of the following functions at the given values.

1. $f(x) = 3x^2 - 2$ at $x = 2$
2. $f(x) = 2x^2 + 1$ at $x = 10$
3. $f(x) = x^2 + 5$ at $x = 3$
4. $f(x) = 5x + 10$ at $x = 10$
5. $f(x) = x^2 + 3x + 5$ at $x = 5$
6. $f(x) = 700x - 2x^2$ at $x = 10$
7. $f(x) = 3x^2 + x + 3$ at $x = 4$

8. $f(x) = x^2 + x - 1$ at $x = -2$
9. $f(x) = 500 - \frac{1}{8}x$ at $x = 8$
10. $f(x) = \frac{30}{4}x - \frac{1}{2}$ at $x = 2$
11. $f(x) = 2x^2 + 3x + 2$ at $x = 5$
12. $f(x) = 300 - 2x^2$ at $x = 10$

13. $f(x) = \dfrac{1}{x}$ at $x = 1$

14. $f(x) = \dfrac{1}{x + 1}$ at $x = 2$

15. Determine f' for $f(x) = \dfrac{1}{x}$.

16. Determine f' for $f(x) = \dfrac{1}{x^2}$.

17. In Exercise 2, $f'(10) = 40$. State four interpretations of this result.
18. In Exercise 5, $f'(5) = 13$. State four interpretations of this result.
19. In Exercise 6, $f'(10) = 660$. State four interpretations of this result.

20. For $f(x) = \begin{cases} x^2, & x \geq 0 \\ -x, & x < 0 \end{cases}$, determine, if it exists, $f'(0)$.

10. Differentiability and continuity

If a function is continuous at a value, does it necessarily have a derivative at the value? The answer is no, that is, continuity does not imply differentiability. A function can be continuous at a value and not have a derivative at the value, as is shown by the function

$$f(x) = \begin{cases} x, & x \geq 0 \\ -x, & x < 0 \end{cases}.$$

In Example 10 (Section 9) it was shown that this function does not have a derivative at 0. It is, however, continuous at 0, since $\displaystyle\lim_{x \to 0} f(x) = f(0) = 0$.

If a function has a derivative at a value, is it necessarily continuous at the value? The answer here is yes, that is, differentiability implies continuity, as we now show.

Theorem. If function f is differentiable at a value a, then it is continuous at a.

Proof. To show that f is continuous at a we must show that $\displaystyle\lim_{x \to a} f(x) = f(a)$. To do this it suffices to show that $\displaystyle\lim_{x \to a} [f(x) - f(a)] = 0$, since to say that $f(x) - f(a)$ approaches 0 is the same as saying that $f(x)$ approaches $f(a)$.

We know from the hypothesis that f is differentiable at a, that is,

$$\lim_{x \to a} \frac{f(x) - f(a)}{x - a} = f'(a).$$

Since we want to demonstrate something about $f(x) - f(a)$ and our hypothesis gives us information about the behavior of $[f(x) - f(a)]/(x - a)$, we must find a way to relate them. Let us observe that for $x \neq a$, the following relationship holds.

$$f(x) - f(a) = \frac{f(x) - f(a)}{(x - a)} \cdot (x - a)$$

By using this relationship and applying the product theorem for limits, we obtain

$$\lim_{x \to a} [f(x) - f(a)] = \lim_{x \to a} \frac{f(x) - f(a)}{x - a} \cdot \lim_{x \to a} (x - a) = f'(a) \cdot 0 = 0.$$

Thus $\lim_{x \to a} [f(x) - f(a)] = 0$, and our theorem is established.

Geometrically speaking, if a function is continuous over an interval, its graph is in one piece over the interval. If a function is differentiable over an interval (that is, has a derivative at each value in the interval), then not only is its graph in one piece, but it is also smooth enough to have a tangent line at each point.

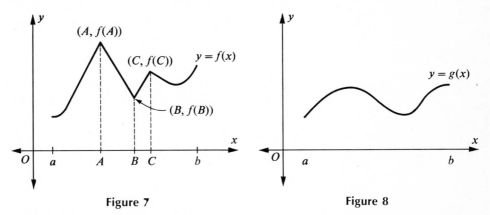

Figure 7   **Figure 8**

The function $y = f(x)$ shown in Figure 7 is continuous at every value in the interval from a to b. $y = f(x)$ is differentiable at every value in that interval, with the exception of values A, B, and C. Geometrically speaking, at points $(A, f(A))$, $(B, f(B))$, and $(C, f(C))$, the corners are too sharp to support a tangent line. The function $y = g(x)$ shown in Figure 8 is differentiable at every value in the interval from a to b.

11. Tools for computing derivatives

Consideration of a number of situations from various backgrounds led us to the concept of derivative. However, the determination of derivatives directly from the definition is only possible for functions with a fairly simple structure. We are thus led to search for theorems which would enable us to compute derivatives in a simple way. Let us begin this search by summarizing the results established in Examples 6–9 (Section 9) on the basis of the definition of derivative.

If $f(x) = x$, then $f'(x) = 1$.
If $f(x) = x^2$, then $f'(x) = 2x$.
If $f(x) = x^3$, then $f'(x) = 3x^2$.
If $f(x) = x^4$, then $f'(x) = 4x^3$.

The pattern exhibited by these examples is the following: If $f(x) = x^n$, then $f'(x) = nx^{n-1}$. We have established this to be the case for $n = 1, 2, 3$, and 4. Is it the case for all positive integers? It can be established on the basis of the definition of derivative that it is. Thus we have the following theorem.

Theorem. If n is a positive integer and f is defined by $f(x) = x^n$, then f has a derivative at each value of x and $f'(x) = nx^{n-1}$.

EXAMPLE 11. Find $f'(x)$ for $f(x) = x^5$, $f(x) = x^{34}$, $f(x) = x^{72}$.

Solution. For $f(x) = x^5$, $f'(x) = 5x^4$ $(n = 5)$.
For $f(x) = x^{34}$, $f'(x) = 34x^{33}$ $(n = 34)$.
For $f(x) = x^{72}$, $f'(x) = 72x^{71}$ $(n = 72)$.

Suppose n is not a positive integer and $f(x) = x^n$; for example, suppose $f(x) = x^{-2}$, or $f(x) = x^{\frac{1}{2}}$. Is it still the case that $f'(x) = nx^{n-1}$? The answer is yes, if n is any rational number and f is defined by $f(x) = x^n$, then f' is given by $f'(x) = nx^{n-1}$.

EXAMPLE 12. For $f(x) = \dfrac{1}{x^2}$, $f(x) = \dfrac{1}{x}$, $f(x) = \sqrt{x}$, $f(x) = x^{-\frac{1}{2}}$, and $f(x) = x^{\frac{2}{3}}$, determine $f'(x)$.

Solution.
$$f(x) = \frac{1}{x^2} = x^{-2} \qquad f'(x) = -2x^{-2-1} = -2x^{-3} = -\frac{2}{x^3}$$

$$f(x) = \frac{1}{x} = x^{-1} \qquad f'(x) = -1x^{-1-1} = -1x^{-2} = -\frac{1}{x^2}$$

$$f(x) = \sqrt{x} = x^{\frac{1}{2}} \qquad f'(x) = \frac{1}{2}x^{\frac{1}{2}-1} = \frac{1}{2}x^{-\frac{1}{2}} = \frac{1}{2\sqrt{x}}$$

$$f(x) = x^{-\frac{1}{2}} \qquad f'(x) = -\frac{1}{2}x^{-\frac{1}{2}-1} = -\frac{1}{2}x^{-\frac{3}{2}}$$

$$f(x) = x^{\frac{2}{3}} \qquad f'(x) = \frac{2}{3}x^{\frac{2}{3}-1} = \frac{2}{3}x^{-\frac{1}{3}}$$

When fractional exponents are involved, the domains of definition of f and f' are not necessarily the same, as shown by $f(x) = \sqrt{x}$ and its derivative, $f'(x) = (1/2)\sqrt{x}$; $f(0) = 0$ while $f'(0)$ is not defined.

In the study of limits we found it helpful to view a function as being made up of component parts put together by the operations of addition, multiplication, and so on, and then to examine the limit behavior of the component parts. If an approach works once, try it again. What is suggested here is that we view a function as being made up of component parts, examine the derivative behavior of the component parts, and try to put the component derivatives together so as to obtain the derivative of the given function. This approach yields a number of theorems which we now state and illustrate. Proofs will be discussed in the next section.

Derivative of a Sum. If $h(x) = f(x) + g(x)$ and $\dfrac{df(x)}{dx}$ and $\dfrac{dg(x)}{dx}$ exist, then

$$\frac{dh(x)}{dx} = \frac{df(x)}{dx} + \frac{dg(x)}{dx}.$$

That is, the derivative of a sum is the sum of the derivatives, provided that the members of the sum have derivatives. More generally this theorem holds for a sum of two or more functions.

EXAMPLE 13.   For $h(x) = x^7 + x^5 + 2$, determine $\dfrac{dh(x)}{dx}$.

Solution.

$$\frac{d(x^7 + x^5 + 2)}{dx} = \frac{d(x^7)}{dx} + \frac{d(x^5)}{dx} + \frac{d(2)}{dx}$$
$$= 7x^6 + 5x^4 + 0$$
$$= 7x^6 + 5x^4$$

Derivative of a Difference. If $h(x) = f(x) - g(x)$ and $\dfrac{df(x)}{dx}$ and $\dfrac{dg(x)}{dx}$ exist, then

$$\frac{dh(x)}{dx} = \frac{df(x)}{dx} - \frac{dg(x)}{dx}.$$

The derivative of a difference is the difference of the derivatives, provided that the members of the difference have derivatives. More generally this theorem holds for a difference of two or more functions.

Exercises

Determine the derivatives of the following functions.

21. $f(x) = x^6$ 22. $f(x) = x^{10}$
23. $f(x) = x^9$ 24. $f(x) = x^{12}$
25. $f(x) = x^{-5}$ 26. $f(x) = x^{-3}$

27. $f(x) = x^{-7}$

28. $f(x) = x^{\frac{1}{5}}$

29. $f(x) = x^7$

30. $f(x) = x^{\frac{3}{4}}$

31. $f(x) = x^{\frac{1}{8}}$

32. $f(x) = x^{\frac{1}{4}}$

33. $f(x) = x^{-\frac{1}{4}}$

34. $f(x) = x^{-\frac{3}{4}}$

35. $f(x) = \dfrac{1}{x^4}$

36. $f(x) = \dfrac{1}{x^5}$

37. $f(x) = \dfrac{1}{\sqrt{x}}$

38. $f(x) = \dfrac{1}{x^{10}}$

39. $f(x) = \dfrac{1}{x^6}$

40. $f(x) = x^3 + x$

41. $f(x) = x^3 + x + 1$

42. $f(x) = x^2 + x - 7$

43. $f(x) = x^4 + x^2 - 10$

44. $f(x) = x^9 + x^5 + 8$

45. $f(x) = x^4 + \dfrac{1}{x} + 3$

46. $f(x) = x^6 + x^2 + \sqrt{x}$

47. $f(x) = x^5 + \dfrac{1}{x^2} + 23$

Derivative of a Product. If $h(x) = f(x) \cdot g(x)$ and $\dfrac{df(x)}{dx}$ and $\dfrac{dg(x)}{dx}$ exist, then

$$\frac{dh(x)}{dx} = f(x)\frac{dg(x)}{dx} + g(x)\frac{df(x)}{dx}.$$

When $f(x)$ is the first member of the product and $g(x)$ is the second member of the product, then the product theorem can be stated in colloquial language as follows.

$$\frac{dh(x)}{dx} = \binom{\text{first}}{\text{function}}\binom{\text{derivative of the}}{\text{second function}} + \binom{\text{second}}{\text{function}}\binom{\text{derivative of the}}{\text{first function}}$$

EXAMPLE 14. For $h(x) = (x^4 + x^3 + 1)(x^2 + x)$, determine $\dfrac{dh(x)}{dx}$.

Solution. By the product theorem,

$$\frac{dh(x)}{dx} = (x^4 + x^3 + 1)\frac{d(x^2 + x)}{dx} + (x^2 + x)\frac{d(x^4 + x^3 + 1)}{dx}.$$

Now,

$$\frac{d(x^2 + x)}{dx} = \frac{d(x^2)}{dx} + \frac{d(x)}{dx} = 2x + 1$$

and

$$\frac{d(x^4 + x^3 + 1)}{dx} = \frac{d(x^4)}{dx} + \frac{d(x^3)}{dx} + \frac{d(1)}{dx} = 4x^3 + 3x^2 + 0 = 4x^3 + 3x^2.$$

Thus we obtain

$$\frac{dh(x)}{dx} = (x^4 + x^3 + 1)(2x + 1) + (x^2 + x)(4x^3 + 3x^2).$$

An important special case of the product theorem arises when $f(x) = c$, that is, when $f(x)$ is a constant function. For this situation we have the following theorem.

Derivative of a Constant Function Times a Function. If $h(x) = c \cdot g(x)$ and $\dfrac{dg(x)}{dx}$ exists, then

$$\frac{dh(x)}{dx} = c\frac{dg(x)}{dx}.$$

That is, the derivative of a constant times a function is equal to the constant times the derivative of the function.

This result follows immediately from the product theorem. By applying the product theorem we obtain

$$\frac{dh(x)}{dx} = c\frac{dg(x)}{dx} + g(x)\frac{d(c)}{dx}.$$

But $\dfrac{d(c)}{dx} = 0$; thus

$$\frac{dh(x)}{dx} = c\frac{dg(x)}{dx}.$$

EXAMPLE 15. Find $\dfrac{dh(x)}{dx}$ for $h(x) = 2x^3$, $h(x) = 5x^7$, $h(x) = 4x^{10}$.

Solution.

$$h(x) = 2x^3 \qquad \frac{dh(x)}{dx} = 2\frac{d(x^3)}{dx} = 2(3x^2) = 6x^2$$

$$h(x) = 5x^7 \qquad \frac{dh(x)}{dx} = 5\frac{d(x^7)}{dx} = 5(7x^6) = 35x^6$$

$$h(x) = 4x^{10} \qquad \frac{dh(x)}{dx} = 4\frac{d(x^{10})}{dx} = 4(10x^9) = 40x^9$$

Derivative of a Quotient. If $h(x) = \dfrac{f(x)}{g(x)}$ and $\dfrac{df(x)}{dx}$ and $\dfrac{dg(x)}{dx}$ exist, then

$$\frac{dh(x)}{dx} = \frac{g(x)\dfrac{df(x)}{dx} - f(x)\dfrac{dg(x)}{dx}}{[g(x)]^2}.$$

In colloquial language we could express this theorem as follows.

$$\frac{dh(x)}{dx} = \frac{(\text{denominator})\begin{pmatrix}\text{derivative of}\\\text{the numerator}\end{pmatrix} - (\text{numerator})\begin{pmatrix}\text{derivative of}\\\text{the denominator}\end{pmatrix}}{(\text{denominator})^2}$$

EXAMPLE 16. For $h(x) = \dfrac{2x^3 + x + 1}{x^2 + 3}$, determine $\dfrac{dh(x)}{dx}$.

Solution. By the quotient theorem,

$$\frac{dh(x)}{dx} = \frac{(x^2 + 3)\dfrac{d(2x^3 + x + 1)}{dx} - (2x^3 + x + 1)\dfrac{d(x^2 + 3)}{dx}}{(x^2 + 3)^2}.$$

Since

$$\frac{d(2x^3 + x + 1)}{dx} = 2\frac{d(x^3)}{dx} + \frac{d(x)}{dx} + \frac{d(1)}{dx} = 6x^2 + 1 + 0 = 6x^2 + 1$$

$$\frac{d(x^2 + 3)}{dx} = \frac{d(x^2)}{dx} + \frac{d(3)}{dx} = 2x + 0 = 2x$$

we obtain

$$\frac{dh(x)}{dx} = \frac{(x^2 + 3)(6x^2 + 1) - (2x^3 + x + 1)(2x)}{(x^2 + 3)^2}.$$

Exercises

Determine the derivatives of the following functions.

48. $f(x) = (x^2 + x + 4)(x^4 + 1)$
49. $f(x) = (x^3 + x^2)(x^4 + x^3 + 5)$
50. $f(x) = 4x^6 + 3x^2 + 2$
51. $f(x) = 3x^5 + 3x^2 + 7$
52. $f(x) = 14x^3 - 7x + 14$
53. $f(x) = \frac{1}{3}x^5 - 8x^3 + \frac{1}{5}x - \frac{1}{2}$
54. $f(x) = 12x^{\frac{1}{2}} - 3x + 23$
55. $f(x) = \frac{1}{4}x^4 + \frac{2}{3}x^2 - 3$
56. $f(x) = (3x^2 - 3x + 1)(4x^3 - 4x)$
57. $f(x) = (7x^3 - \frac{1}{2}x)(3x^6 - 4x + 12)$
58. $f(x) = (4x^3 + 3x^2 - 2x)(5x^4 - 6x^2 - 19)$
59. $f(x) = (7x^6 + 4x^3 - 13)(6x^3 - 9x^2 - 13x)$

60. $f(x) = (3x^6 + 4x^5 + 1)\left(4x^2 + \dfrac{3}{x^2}\right)$

61. $f(x) = \dfrac{3x + 1}{2x - 5}$

62. $f(x) = \dfrac{2x^3 - 4}{3x + 1}$

63. $f(x) = \dfrac{12x^2 + 4}{3x^2 + 1}$

64. $f(x) = \dfrac{3x^3 + 9}{-4x^6 + 10}$

65. $f(x) = \dfrac{x^2 + 1}{2x^3 + 3x + 1}$

66. $f(x) = \dfrac{3x^3 + 2x + 1}{x^4 + 2x^2 - 7}$

67. $f(x) = \dfrac{5x^6 - 13x^4 - 43}{9x^5 - 12x}$

68. $f(x) = \dfrac{12x^9 - 7x^5 - 4x + 15}{6x^7 - 9x^4 + 18}$

The chain rule

Consider the problem of determining the derivative of

$$y = (x^3 + x)^2.$$

This problem can be handled by multiplying $(x^3 + x)$ by itself, obtaining

$$y = x^6 + 2x^4 + x^2.$$

Thus

$$\frac{dy}{dx} = 6x^5 + 8x^3 + 2x.$$

If on the other hand we were required to find the derivative of

$$y = (x^3 + x)^{100} \quad \text{or} \quad y = (x^3 + x)^{\frac{1}{2}}$$

we would find the tools developed thus far inadequate. A new approach is needed. In developing it, let us return to

$$y = (x^3 + x)^2.$$

The problem here is with $(x^3 + x)$, which is squared. If we were dealing with $y = u^2$ the problem would be simple: $\dfrac{dy}{du} = 2u$. Suppose we introduce an auxiliary function $u = x^3 + x$. Then from

$$y = \underbrace{(x^3 + x)}_{u}{}^2$$

we obtain a second auxiliary function, $y = u^2$. There are now three functions on the scene; the original function

$$y = (x^3 + x)^2$$

in which y is a function of x, and the two auxiliary functions

$$u = x^3 + x, \qquad y = u^2.$$

Remarkably, the derivatives of these functions are related in a strikingly simple manner. The derivative of the original function is equal to the product of the derivatives of the auxiliary functions.

$$\underbrace{\frac{dy}{dx}}_{\substack{\text{derivative} \\ \text{of the} \\ \text{original} \\ \text{function}}} = \underbrace{\frac{dy}{du} \cdot \frac{du}{dx}}_{\substack{\text{product of the} \\ \text{derivatives of} \\ \text{the auxiliary} \\ \text{functions}}}.$$

This result is called the *composite-function theorem* or the *chain rule*. For the problem under discussion,

$$y = u^2, \qquad \frac{dy}{du} = 2u$$

$$u = x^3 + x, \qquad \frac{du}{dx} = 3x^2 + 1.$$

Thus from the chain rule we obtain

$$\frac{dy}{dx} = \frac{dy}{du} \cdot \frac{du}{dx} = 2u(3x^2 + 1).$$

Replacing u by $x^3 + x$ yields

$$\frac{dy}{dx} = 2(x^3 + x)(3x^2 + 1).$$

By multiplying we obtain

$$\frac{dy}{dx} = 6x^5 + 8x^3 + 2x$$

which is also the result we obtain by multiplying $(x^3 + x)$ by itself, then determining the derivative of each term.

EXAMPLE 17. For $y = (x^3 + x)^{100}$, determine $\dfrac{dy}{dx}$.

Solution. $y = (x^3 + x)^{100}$; let $u = x^3 + x$ so that we obtain as our second auxiliary function $y = u^{100}$. Our basic tool is the chain rule,

$$\frac{dy}{dx} = \frac{dy}{du} \cdot \frac{du}{dx}.$$

Since $y = u^{100}$,

$$\frac{dy}{du} = 100u^{99}.$$

Since $u = x^3 + x$,

$$\frac{du}{dx} = 3x^2 + 1.$$

Thus by the chain rule we have

$$\frac{dy}{dx} = \frac{dy}{du} \cdot \frac{du}{dx} = 100u^{99}(3x^2 + 1).$$

Replacing u by $x^3 + x$ yields

$$\frac{dy}{dx} = 100(x^3 + x)^{99}(3x^2 + 1).$$

EXAMPLE 18. For $y = (x^3 + x)^{\frac{1}{2}}$, determine $\dfrac{dy}{dx}$.

Solution. $y = (x^3 + x)^{\frac{1}{2}}$; let $u = x^3 + x$ so that we obtain as our second auxiliary function $y = u^{\frac{1}{2}}$. Our basic tool is the chain rule.

$$\frac{dy}{dx} = \frac{dy}{du} \cdot \frac{du}{dx}$$

Since $y = u^{\frac{1}{2}}$,

$$\frac{dy}{du} = \frac{1}{2} u^{-\frac{1}{2}}.$$

Since $u = x^3 + x$,

$$\frac{du}{dx} = 3x^2 + 1.$$

Thus by the chain rule we have

$$\frac{dy}{dx} = \frac{dy}{du} \cdot \frac{du}{dx} = \frac{1}{2} u^{-\frac{1}{2}}(3x^2 + 1).$$

Replacing u by $x^3 + x$ yields

$$\frac{dy}{dx} = \frac{1}{2}(x^3 + x)^{-\frac{1}{2}}(3x^2 + 1) = \frac{3x^2 + 1}{2\sqrt{x^3 + x}}.$$

EXAMPLE 19. For $y = x^2(x^3 + x)^{100}$, determine $\dfrac{dy}{dx}$.

Solution. By the product theorem we obtain

$$\frac{dy}{dx} = x^2 \frac{d(x^3 + x)^{100}}{dx} + (x^3 + x)^{100} \frac{d(x^2)}{dx}.$$

Now, $\dfrac{d(x^2)}{dx} = 2x$. Let $y_1 = (x^3 + x)^{100}$. By the chain rule (see Example 17) we have

$$\frac{dy_1}{dx} = 100(x^3 + x)^{99}(3x^2 + 1).$$

Substitution of these results into the expression for $\dfrac{dy}{dx}$ yields

$$\frac{dy}{dx} = 100x^2(3x^2 + 1)(x^3 + x)^{99} + 2x(x^3 + x)^{100}.$$

Exercises

Determine the derivatives of the following functions.

69. $f(x) = (8x^4 + 2x + 1)^{10}$
70. $f(x) = (3x^3 + 2x^2 + 8)^6$
71. $f(x) = (4x^2 + 2x + 3)^{\frac{1}{2}}$
72. $f(x) = (x^6 + 2x + 2)^{\frac{1}{3}}$
73. $f(x) = (x^4 + 3x - 8)^{\frac{1}{4}}$
74. $f(x) = 3x^4(3x^2 + 1)^5$
75. $f(x) = 6x^2(x^3 + 2x - 10)^{\frac{1}{2}}$
76. $f(x) = \dfrac{(x^3 + 5)^3}{x^2 - 7}$
77. $f(x) = \dfrac{3x^2 + 2x - 9}{(4x^3 - 3x)^5}$
78. $f(x) = (4x^5 + 2x - 8)^4$
79. $f(x) = \sqrt{3x^2 + 2}$
80. $f(x) = \sqrt[3]{4x^3 + 2x - 1}$
81. $f(x) = \sqrt{5x^2 - 12}$
82. $f(x) = 5x^6\sqrt{7x^5 - 13}$
83. $f(x) = \dfrac{7x}{\sqrt{2x + 1}}$
84. $f(x) = \dfrac{3x^2}{\sqrt{3x - 7}}$

12. Proofs of some theorems on derivatives*

In the previous section we stated and illustrated a number of theorems for derivatives. Here we will see how some of these theorems are proved.

Derivative of a Sum. If $h(x) = f(x) + g(x)$ and $f'(a)$ and $g'(a)$ exist, then $h'(a)$ exists and

$$h'(a) = f'(a) + g'(a).$$

Proof. We must show that $h'(a)$, which by definition is

$$\lim_{x \to a} \frac{h(x) - h(a)}{x - a}$$

is equal to $f'(a) + g'(a)$.

Let us begin by observing that $h(x) = f(x) + g(x)$ and $h(a) = f(a) + g(a)$. Substitution and elementary algebra yield the following results.

$$\frac{h(x) - h(a)}{x - a} = \frac{f(x) + g(x) - [f(a) + g(a)]}{x - a}$$

$$= \frac{f(x) - f(a) + g(x) - g(a)}{x - a}$$

$$= \frac{f(x) - f(a)}{x - a} + \frac{g(x) - g(a)}{x - a}$$

By hypothesis $f'(a)$ and $g'(a)$ exist, that is,

$$\lim_{x \to a} \frac{f(x) - f(a)}{x - a} = f'(a) \quad \text{and} \quad \lim_{x \to a} \frac{g(x) - g(a)}{x - a} = g'(a).$$

Thus by the limit theorem for sums we obtain

$$h'(a) = \lim_{x \to a} \frac{h(x) - h(a)}{x - a} = \lim_{x \to a} \frac{f(x) - f(a)}{x - a} + \lim_{x \to a} \frac{g(x) - g(a)}{x - a}$$

$$= f'(a) + g'(a).$$

Derivative of a Product. If $h(x) = f(x) \cdot g(x)$ and $f'(a)$ and $g'(a)$ exist, then $h'(a)$ exists and

$$h'(a) = f(a) \cdot g'(a) + g(a) \cdot f'(a).$$

Proof. We must show that $h'(a)$, that is,

$$\lim_{x \to a} \frac{h(x) - h(a)}{x - a}$$

is equal to $f(a) \cdot g'(a) + g(a) \cdot f'(a)$.

*This section may be omitted without disturbing continuity.

Since $h(x) = f(x) \cdot g(x)$ and $h(a) = f(a) \cdot g(a)$, we have

$$\frac{h(x) - h(a)}{x - a} = \frac{f(x) \cdot g(x) - f(a) \cdot g(a)}{x - a}. \tag{1}$$

Our problem is to express the right member of equation (1) in terms of

$$\frac{f(x) - f(a)}{x - a} \quad \text{and} \quad \frac{g(x) - g(a)}{x - a}.$$

To do this we will employ a stratagem that mathematicians have borrowed from politicians: add and subtract the same quantity to the numerator of the right member of equation (1). This is permissible because the total contribution is zero—that is, nothing is changed. (In political circles this is called giving with one hand and taking with the other.)

Adding and subtracting $f(a) \cdot g(x)$ to the numerator of the right member of (1) yields

$$\frac{h(x) - h(a)}{x - a} = \frac{f(x) \cdot g(x) - f(a) \cdot g(x) + f(a) \cdot g(x) - f(a) \cdot g(a)}{x - a}.$$

By factoring we obtain

$$\frac{h(x) - h(a)}{x - a} = \frac{g(x)[f(x) - f(a)] + f(a)[g(x) - g(a)]}{x - a}$$

$$= g(x)\frac{f(x) - f(a)}{x - a} + f(a)\frac{g(x) - g(a)}{x - a}. \tag{2}$$

Since g is differentiable at a, it is continuous at a. Thus

$$\lim_{x \to a} g(x) = g(a).$$

Also, $\lim_{x \to a} f(a) = f(a)$ since $f(a)$ is a constant. By hypothesis,

$$\lim_{x \to a} \frac{f(x) - f(a)}{x - a} = f'(a) \quad \text{and} \quad \lim_{x \to a} \frac{g(x) - g(a)}{x - a} = g'(a).$$

Thus by the limit theorems for sums and for products applied to (2) we obtain

$$\lim_{x \to a} \frac{h(x) - h(a)}{x - a} = [\lim_{x \to a} g(x)] \cdot \lim_{x \to a} \frac{f(x) - f(a)}{x - a} + [\lim_{x \to a} f(a)] \cdot \lim_{x \to a} \frac{g(x) - g(a)}{x - a}.$$

Therefore

$$h'(a) = g(a) \cdot f'(a) + f(a) \cdot g'(a)$$

or equivalently

$$h'(a) = f(a) \cdot g'(a) + g(a) \cdot f'(a).$$

Derivative of a Quotient. If $h(x) = \dfrac{f(x)}{g(x)}$ and $f'(a)$ and $g'(a)$ exist, then $h'(a)$ exists and

$$h'(a) = \frac{g(a) \cdot f'(a) - f(a) \cdot g'(a)}{[g(a)]^2}.$$

Proof. Our task is to show that $h'(a)$, that is,

$$\underset{x \to a}{\text{limit}} \frac{h(x) - h(a)}{x - a}$$

is equal to

$$\frac{g(a) \cdot f'(a) - f(a) \cdot g'(a)}{[g(a)]^2}.$$

Since $h(x) = f(x)/g(x)$ and $h(a) = f(a)/g(a)$, we have

$$h(x) - h(a) = \frac{f(x)}{g(x)} - \frac{f(a)}{g(a)}.$$

By using $g(x) \cdot g(a)$ as a common denominator and combining terms, we obtain

$$h(x) - h(a) = \frac{f(x) \cdot g(a) - g(x) \cdot f(a)}{g(x) \cdot g(a)}. \tag{3}$$

Adding and subtracting $f(a) \cdot g(a)$ to the numerator of the right member of equation (3) yields

$$h(x) - h(a) = \frac{f(x) \cdot g(a) - f(a) \cdot g(a) - g(x) \cdot f(a) + f(a) \cdot g(a)}{g(x) \cdot g(a)}. \tag{4}$$

By factoring we obtain

$$h(x) - h(a) = \frac{g(a)[f(x) - f(a)] - f(a)[g(x) - g(a)]}{g(x) \cdot g(a)}. \tag{5}$$

Multiplying both members of (5) by $1/(x - a)$ and using basic algebra yields

$$\frac{h(x) - h(a)}{x - a} = \frac{1}{g(x) \cdot g(a)} \left[g(a) \left(\frac{f(x) - f(a)}{x - a} \right) - f(a) \left(\frac{g(x) - g(a)}{x - a} \right) \right]. \tag{6}$$

Preparatory to using our limit theorems, let us examine the limit behavior of the component parts of (6). Since g is differentiable at a, it is continuous at a. Thus

$$\underset{x \to a}{\text{limit}}\, g(x) = g(a).$$

Since $f(a)$ and $g(a)$ are constants,

$$\lim_{x \to a} f(a) = f(a) \qquad \text{and} \qquad \lim_{x \to a} g(a) = g(a).$$

By hypothesis,

$$\lim_{x \to a} \frac{f(x) - f(a)}{x - a} = f'(a) \quad \text{and} \quad \lim_{x \to a} \frac{g(x) - g(a)}{x - a} = g'(a).$$

Thus from (6), the above observations on the limit behavior of the component parts of (6), and our limit theorems, we obtain

$$h'(a) = \lim_{x \to a} \frac{h(x) - h(a)}{x - a} = \frac{1}{g(a) \cdot g(a)} [g(a) \cdot f'(a) - f(a) \cdot g'(a)]$$

$$= \frac{g(a) \cdot f'(a) - f(a) \cdot g'(a)}{[g(a)]^2}.$$

Exercise

85. Prove the theorem for the derivative of a difference: If $h(x) = f(x) - g(x)$ and $f'(a)$ and $g'(a)$ exist, then $h'(a)$ exists and $h'(a) = f'(a) - g'(a)$.

13. The inverse of a function and its derivative

The function

$$y = 2x + 5$$

defines y as a function of x. x is the independent variable and y is the dependent variable; values for x are chosen and corresponding values for y are determined. But in this situation x can also be defined as a function of y. Solving for x in terms of y yields

$$x = \tfrac{1}{2}y - \tfrac{5}{2}.$$

Here y is the independent variable and x is the dependent variable; values for y are chosen and corresponding values for x are determined. The functions $y = 2x + 5$ and $x = \tfrac{1}{2}y - \tfrac{5}{2}$ are said to be inverses, and each one is said to be the inverse of the other. More generally, if $y = f(x)$ defines y as a function of x and x can be expressed as a function of y, $x = g(y)$, then the functions $y = f(x)$ and $x = g(y)$ are said to be *inverses,* and each one is said to be the inverse of the other.

In many situations two variables x and y are singled out for study, and it is as desirable to express y as a function of x as it is to express x as a function of y. Such is the case, for example, in the study in economics of the relationship between market demand for a commodity and the price of the commodity. If $x = g(y)$ expresses the quantity x of a commodity purchased by the market per unit time as a function of the price y of the commodity, then the inverse $y = f(x)$ expresses the price y of the commodity as a function of the quantity purchased by the market.

Suppose

$$x = -5y + 30$$

describes the quantity of pork purchased in millions of pounds per month as a function of the price of pork in dollars per pound. Then the inverse function

$$y = -\tfrac{1}{5}x + 6$$

describes the price of pork in dollars per pound as a function of the quantity of pork in millions of pounds purchased per month by the market.

Let us observe that both $y = 2x + 5$ and its inverse $x = \tfrac{1}{2}y - \tfrac{5}{2}$ have derivatives,

$$\frac{dy}{dx} = 2 \qquad \frac{dx}{dy} = \frac{1}{2}$$

and these derivatives are related by

$$\frac{dx}{dy} = \frac{1}{\dfrac{dy}{dx}}.$$

Such a relationship between derivatives of inverse functions holds in general. If $y = f(x)$ and $x = g(y)$ are inverses and $\dfrac{dy}{dx}$ and $\dfrac{dx}{dy}$ exist, then in any interval where $\dfrac{dy}{dx} \neq 0$

$$\frac{dx}{dy} = \frac{1}{\dfrac{dy}{dx}}.$$

By means of this theorem, the derivative of the inverse function can be found without determining the inverse function explicitly.

To further illustrate, consider $y = x^3$ and its inverse $x = y^{\frac{1}{3}}$. From our theorem we have

$$\frac{dx}{dy} = \frac{1}{\dfrac{dy}{dx}} = \frac{1}{3x^2} = \frac{1}{3(y^{\frac{1}{3}})^2} = \frac{1}{3} \cdot \frac{1}{y^{\frac{2}{3}}} = \frac{1}{3}y^{-\frac{2}{3}}.$$

EXAMPLE 20. For $x = y^{\frac{1}{2}}$, $y > 0$, find the derivative of its inverse.

Solution. If $y = f(x)$ is the inverse of $x = y^{\frac{1}{2}}$, then

$$\frac{dy}{dx} = \frac{1}{\dfrac{dx}{dy}} = \frac{1}{\tfrac{1}{2}y^{-\frac{1}{2}}} = 2y^{\frac{1}{2}}.$$

If $y = f(x)$ were not known, we would stop here. But since we know that the inverse of $x = y^{\frac{1}{2}}$, $y > 0$, is $y = x^2$, $x > 0$, let us carry this one step further and express $\dfrac{dy}{dx}$ in terms of x by substituting x^2 for y. We obtain

$$\frac{dy}{dx} = 2y^{\frac{1}{2}} = 2(x^2)^{\frac{1}{2}} = 2x.$$

Exercises

Find the inverse functions of the following functions.

86. $y = x + 7$
87. $y = 3x - 10$
88. $x = 8y^3$
89. $x = (y + 1)^3$

90. $y = \dfrac{x + 1}{x}$
91. $y = \log_{10} x$

92. $y = \log_b x$

Find the derivatives of the inverses of the following functions.

93. $y = x + 7$
94. $y = 3x - 10$
95. $x = 8y^3$
96. $x = (y + 1)^3$

97. $x = \dfrac{4}{y}$
98. $y = x^3 + 64$

99. $y = \dfrac{x + 1}{x}$

14. Exponential functions and their derivatives

The functions

$$y = \left(\frac{1}{8}\right)^x, \quad y = 2^x, \quad y = 10^x, \quad y = e^x$$

where $e = 2.71828$ correct to five decimal places, and e is defined as

$$e = \operatorname*{limit}_{n \to \infty} \left(1 + \frac{1}{n}\right)^n$$

(see Section 7), illustrate exponential functions of the most basic type. All of these functions are of the form

$$y = b^x$$

where b is a positive constant and $b \neq 1$. These functions, their derivatives, and their related functions occur frequently in mathematics and its applications. The general curve for $y = b^x$ for $0 < b < 1$ is shown in Figure 9, and in Figure 10 the general curve for $y = b^x$ for $b > 1$ is shown.

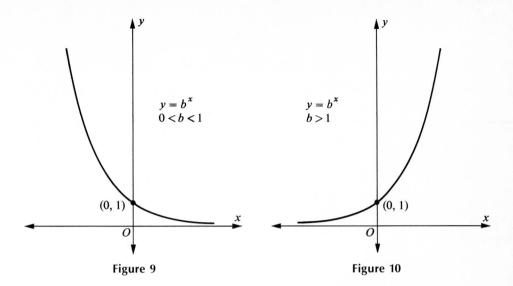

Figure 9   **Figure 10**

Exponential functions, their related functions, and their derivatives appear in a number of situations in the worlds of biology, business, economics, and psychology. We have already observed in Section 7 that the exponential function

$$a(x) = A\, e^{rx}$$

expresses the asset value that an initial amount of money A will grow to in x years if invested at rate r compounded continuously. Also, we saw that

$$b(x) = A\, e^{-rx}$$

expresses the amount that must initially be invested at rate r compounded continuously if at the end of x years amount A is to be made available. The graphs of these functions for $x \geq 0$ are shown in Figures 11 and 12 on the next page. In both cases, A and r are positive constants.

In his learning-theory studies, C. L. Hull* was led to the function

$$h(x) = 100(1 - e^{-ax})$$

where a is a positive constant, to describe habit strength in terms of repetitions. The graph of this function for $x \geq 0$ is shown in Figure 13 on the next page.

The Gompertz curves,† defined by functions of the form

$$f(x) = k\, A^{(b^x)}$$

where k, A, and b are positive constants and A and b are less than one, have been used by biologists and psychologists in the study of growth situations and by the

---

*C. L. Hull, *Principles of Behavior* (New York: Appleton-Century-Crofts, 1943).

†For discussion of the use of Gompertz curves in time-series analysis, see Freund and Williams, *Modern Business Statistics,* rev. ed. (Englewood Cliffs, New Jersey: Prentice-Hall, Inc., 1969), pp. 420–421.

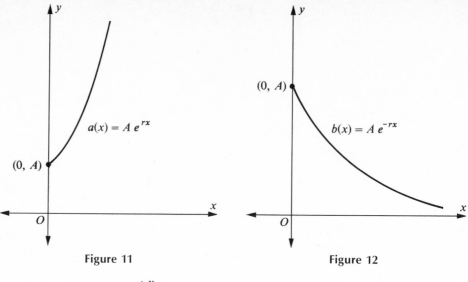

Figure 11   Figure 12

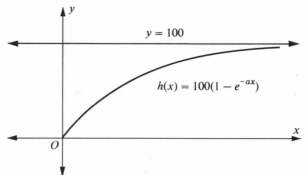

Figure 13

National Industrial Conference Board in its studies of growth patterns of industries, cities, and states. The graph of a Gompertz curve is shown in Figure 14.

As far as derivatives are concerned, it can be shown that if $y = b^x$, then

$$\frac{dy}{dx} = (\log_e b)b^x.*$$

If $y = \left(\frac{1}{8}\right)^x$, $\frac{dy}{dx} = \left(\log_e \frac{1}{8}\right)\left(\frac{1}{8}\right)^x$.

If $y = 2^x$, $\frac{dy}{dx} = (\log_e 2)2^x$.

If $y = 10^x$, $\frac{dy}{dx} = (\log_e 10)10^x$.

---

*For a review of logarithms, see Appendix 3.

If $y = e^x$, $\quad \dfrac{dy}{dx} = (\log_e e)e^x = e^x.$

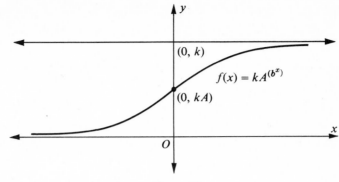

Figure 14

EXAMPLE 21. In Section 7 we saw that if $1000 is invested at 5% compounded continuously, then its asset value in x years is given by the function

$$a(x) = 1000\, e^{0.05x}.$$

Determine $\dfrac{d[a(x)]}{dx}$.

Solution.

$$\frac{d[a(x)]}{dx} = 1000\, \frac{d(e^{0.05x})}{dx} \tag{1}$$

Let $y_1 = e^{0.05x}$. To find $\dfrac{dy_1}{dx}$ we must use the chain rule. Let $u = 0.05x$; then we obtain $y_1 = e^u$ as our second auxiliary function. From the chain rule we have

$$\frac{dy_1}{dx} = \frac{dy_1}{du} \cdot \frac{du}{dx}.$$

Since $y_1 = e^u$,

$$\frac{dy_1}{du} = e^u.$$

Since $u = 0.05x$,

$$\frac{du}{dx} = 0.05.$$

Thus from the chain rule we obtain

$$\frac{dy_1}{dx} = \frac{dy_1}{du} \cdot \frac{du}{dx} = e^u(0.05) = (0.05)e^u.$$

Replacing u by $0.05x$ yields

$$\frac{dy_1}{dx} = (0.05)e^{0.05x}.$$

By substituting this result into equation (1) we obtain

$$\frac{d[a(x)]}{dx} = (1000)(0.05)e^{0.05x} = 50\ e^{0.05x}.$$

EXAMPLE 22. For $y = e^{2x+1}$, determine $\dfrac{dy}{dx}$.

Solution. $y = e^{2x+1}$; let $u = 2x + 1$, so that we obtain $y = e^u$ as our second auxiliary function. Our basic tool is the chain rule.

$$\frac{dy}{dx} = \frac{dy}{du} \cdot \frac{du}{dx}$$

Since $y = e^u$,

$$\frac{dy}{du} = e^u.$$

Since $u = 2x + 1$,

$$\frac{du}{dx} = 2.$$

Thus by the chain rule we obtain

$$\frac{dy}{dx} = \frac{dy}{du} \cdot \frac{du}{dx} = e^u(2).$$

Replacing u by $2x + 1$ yields

$$\frac{dy}{dx} = 2\ e^{2x+1}.$$

EXAMPLE 23. For $y = x^2 \cdot 2^{8x}$, determine $\dfrac{dy}{dx}$.

Solution. By the product theorem

$$\frac{dy}{dx} = x^2 \frac{d(2^{8x})}{dx} + 2^{8x} \frac{d(x^2)}{dx}. \tag{1}$$

Now $\dfrac{d(x^2)}{dx} = 2x$. From the chain rule we obtain

$$\frac{d(2^{8x})}{dx} = 8(\log_e 2)2^{8x}.$$

Substituting these results into equation (1) yields

$$\frac{dy}{dx} = 8x^2(\log_e 2)2^{8x} + (2x)2^{8x}.$$

EXAMPLE 24.   A class of revenue functions studied in economics is of the form

$$R(x) = Ax\,b^{cx}$$

where b is a constant (a positive number between 0 and 1) and A and c are constants. For $A = 950$, $b = \frac{1}{8}$, and $c = 7$, we obtain

$$R(x) = 950x\left(\frac{1}{8}\right)^{7x}$$

as the revenue function for a producer supplying x units per unit time of some commodity to the market. Find the marginal revenue for an output of x units.

Solution.   The marginal revenue for an output of x units is, by definition, $\dfrac{d[R(x)]}{dx}$. By the product theorem we have

$$\frac{d[R(x)]}{dx} = 950x\,\frac{d\left(\frac{1}{8}\right)^{7x}}{dx} + \left(\frac{1}{8}\right)^{7x}\frac{d(950x)}{dx}. \tag{1}$$

Now $\dfrac{d(950x)}{dx} = 950$. From the chain rule we obtain

$$\frac{d\left(\frac{1}{8}\right)^{7x}}{dx} = 7\left(\log_e \frac{1}{8}\right)\left(\frac{1}{8}\right)^{7x}.$$

Thus by substituting these results into equation (1) we obtain

$$\frac{d[R(x)]}{dx} = 6650x\left(\log_e \frac{1}{8}\right)\left(\frac{1}{8}\right)^{7x} + 950\left(\frac{1}{8}\right)^{7x}$$

as the marginal revenue for an output of x units.

Exercises

100. By plotting a number of points and connecting them with a smooth curve, sketch the graphs of $y = 2^x$ and $y = 3^x$.

101. Consider the following growth situation. A population at a certain time, which we will call time $t = 0$ since it is the starting point of our observations, has A individuals. At time $t = 1$, each member of the population gives rise to $b - 1$ additional members, so that the number of individuals in the population at time $t = 1$ is

$$A + A(b - 1) = A + Ab - A = Ab.$$

At time $t = 2$, each member of the population gives rise to $b - 1$ additional members, so that the number of individuals in the population at time $t = 2$ is

$$Ab + Ab(b - 1) = Ab + Ab^2 - Ab = Ab^2.$$

And so on. By studying this growth pattern, derive the function which describes the number of individuals in the population at any time t.

102. A newly formed university is beginning its operations with an administrative staff of 3 deans. At the end of each year, each person hires two assistants.
  a. Using the results obtained in Exercise 101, state the function which describes the number of administrative personnel at the end of t years.
  b. How many administrative personnel will the university have at the end of five years?
103. If each bacterium in a medium consisting of 10 E. Coli organisms subdivides at the end of each second, state the function which describes the number of bacteria that will be in the medium at the end of t seconds. Determine the number of bacteria that will be in the medium at the end of six seconds.

Determine the derivatives of the following functions.

104. $y = 3^x$

105. $y = 12^x$

106. $y = 4(2^x) + 3x + 1$

107. $y = 3x^2 - 4^x$

108. $y = 3 e^x - 5x^3$

109. $y = x^3 e^x$

110. $y = 4x^2 e^x$

111. $y = x^2 2^x$

112. $y = e^{4x-1}$

113. $y = 3 e^{2x-3}$

114. $y = 2000 e^{0.12x}$

115. $y = A e^{rx}$, where A and r are positive constants

116. $y = 5000 e^{-0.3x}$

117. $y = A e^{-rx}$, where A and r are positive constants

118. $y = 100(1 - e^{-ax})$, where $a > 0$ is constant (C. L. Hull's learning-theory function)

119. $y = x^2 e^{-x}$

120. $y = \dfrac{3 e^x}{x^2 + 1}$

121. $y = \dfrac{3x^2 + 7}{e^x + 4}$

122. $y = (3x^4 + 2x)2^x$

123. $y = \dfrac{e^{-x}}{3x + 7}$

124. $y = e^{x^3+2}$

125. $y = \dfrac{1}{\sqrt{2\pi}} e^{-x^2/2}$

126. $y = k A^{(b^x)}$ (Gompertz curve functions)

127. $y = 1000 e^{\sqrt{x}-rx}$, where r is a positive constant

128. $y = 500 e^{0.5\sqrt{x}-0.10x}$

15. Logarithmic functions and their derivatives

Let us recall that y is said to be the *logarithm* of x with respect to base b, expressed by

$$y = \log_b x$$

if b raised to the power y is x, that is, if $b^y = x$. Thus $\log_{10} 100 = 2$ since $10^2 = 100$; $\log_3 81 = 4$ since $3^4 = 81$; $\log_{10} 1 = 0$ since $10^0 = 1$; $\log_e e = 1$ since $e^1 = e$. $y = \log_b x$ is defined whenever x is positive and base b is positive, $b \neq 1$. The graph of $y = \log_b x$ for $0 < b < 1$ is shown in Figure 15, and the graph of $y = \log_b x$ for $b > 1$ is shown in Figure 16.

One of the most interesting situations in which a logarithmic function makes its appearance is in connection with the potassium-argon clock. In the late 1930s it was hypothesized that Potassium 40, in the course of undergoing radioactive decay, is transformed into Argon 40. This theory was confirmed by experimental findings in the early 1940s, and by 1948 the foundations for the successful exploitation of the nature of this process for solving geological-dating problems had been laid. By the mid-1950s, potassium-argon dating was a well-established dating technique. In contrast to the Carbon 14 method and certain other dating techniques which measure the disappearance of radioactive atoms, the potassium-argon clock is an accumulation clock, relying on the accumulation of Argon 40 atoms in the rock in question. The potassium-argon age function is the logarithmic function

$$t = (1.885 \times 10^9) \log_e [9.068x + 1]$$

where t is age in years and x is the ratio of the amount of Argon 40 to the amount of Potassium 40 in the rock today. The potassium-argon dating method can be used on rocks as young as 10,000 years as well as the oldest rocks known.

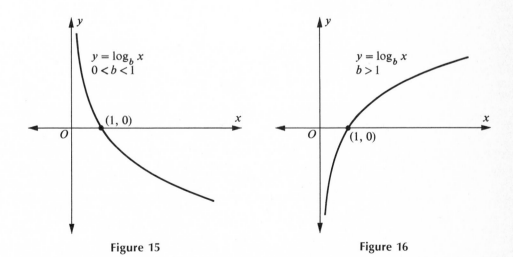

Figure 15   Figure 16

One of the most exciting archeological finds of recent years was the discovery by Dr. and Mrs. Louis S. B. Leakey in 1959 of a fossil hominid skull (which they called *Zinjanthropus boisei*) at Olduvai Gorge in Tanzania. Potassium-argon dating of the volcanic material associated with the hominid remains of Oldavai Gorge played a key role in establishing without question that the Zinjanthropus remains are about 1.75 million years old.*

Since $y = \log_b x$ means that $x = b^y$, it follows by definition that these functions are inverses. Both of these functions have derivatives, so we can determine $\dfrac{dy}{dx}$ by using our theorem for the derivative of an inverse (Section 13). From this theorem we have

$$\frac{dy}{dx} = \frac{1}{\dfrac{dx}{dy}} = \frac{1}{(\log_e b)b^y}.$$

Replacing b^y by x yields

$$\frac{dy}{dx} = \frac{1}{(\log_e b)} \cdot \frac{1}{x}.$$

In algebra (see Appendix 3) it is shown that $\log_a b = 1/\log_b a$. Applying this result here yields

$$\frac{dy}{dx} = \frac{1}{(\log_e b)} \cdot \frac{1}{x} = (\log_b e) \cdot \frac{1}{x}.$$

Thus, in summary, we have the following theorem.

Theorem. If $y = \log_b x$, then $\dfrac{dy}{dx} = (\log_b e) \cdot \dfrac{1}{x}$.

If $y = \log_{10} x$, $\qquad \dfrac{dy}{dx} = (\log_{10} e) \cdot \dfrac{1}{x}$.

If $y = \log_2 x$, $\qquad \dfrac{dy}{dx} = (\log_2 e) \cdot \dfrac{1}{x}$.

If $y = \log_e x$, $\qquad \dfrac{dy}{dx} = (\log_e e) \cdot \dfrac{1}{x} = \dfrac{1}{x}$.

Logarithms with respect to base e are of particular importance in mathematics and its applications and are often referred to as *natural logarithms*. The logarithmic function $y = \log_e x$ is often expressed by the notation

$$y = \ln x.$$

---

*For further discussion of potassium-argon dating, see G. B. Dalrymple and M. A. Lanphere, *Potassium-Argon Dating* (San Francisco: W. H. Freeman, 1969).

EXAMPLE 25. For $y = x^2 \ln x$, find $\dfrac{dy}{dx}$.

Solution. By the product theorem we obtain

$$\frac{dy}{dx} = x^2 \frac{d(\ln x)}{dx} + (\ln x) \frac{d(x^2)}{dx}.$$

Since $\dfrac{d(\ln x)}{dx} = \dfrac{1}{x}$ and $\dfrac{d(x^2)}{dx} = 2x$, we have

$$\frac{dy}{dx} = x^2 \cdot \frac{1}{x} + (\ln x)2x = x + 2x(\ln x).$$

EXAMPLE 26. For $y = \ln (x^2 + 1)$, determine $\dfrac{dy}{dx}$.

Solution. To determine $\dfrac{dy}{dx}$ we must use the chain rule. $y = \ln (x^2 + 1)$; if we let $u = x^2 + 1$, then we obtain $y = \ln u$ as our second auxiliary function. Again, our basic tool is the chain rule.

$$\frac{dy}{dx} = \frac{dy}{du} \cdot \frac{du}{dx}$$

Since $y = \ln u$,

$$\frac{dy}{du} = \frac{1}{u}.$$

Since $u = x^2 + 1$,

$$\frac{du}{dx} = 2x.$$

Thus by the chain rule we obtain

$$\frac{dy}{dx} = \frac{dy}{du} \cdot \frac{du}{dx} = \frac{1}{u}(2x).$$

Replacing u by $x^2 + 1$ yields

$$\frac{dy}{dx} = \frac{2x}{x^2 + 1}.$$

Exercises

Determine the derivatives of the following functions.

129. $y = \log_3 x$ 130. $y = \log_5 x$

131. $y = 3x(\ln x)$ 132. $y = 4x^3(\ln x)$

133. $y = \dfrac{3x^2 + 5}{\ln x}$ 134. $y = (5x^3 + 4x + 7)\ln x$

135. $y = \ln(3x^4 + 2x)$ 136. $y = e^x \ln x$

137. $y = \dfrac{\ln x}{e^x}$ 138. $y = \ln(5x^2 - 19)$

139. $y = \log_2(6x^5 - 4)$ 140. $y = \log_3(5x + 1)$

141. $y = 2^x \ln x$ 142. $y = e^{-x} \ln(x^2 + 1)$

143. $y = \dfrac{2^x + x^2}{\ln x}$ 144. $y = (\ln x)^3$

16. Implicit differentiation*

A function is said to be defined *explicitly* if its dependent variable is expressed directly in terms of its independent variable. $y = 2x + 5$ and $y = 2x^2 + 3$ illustrate explicitly defined functions.

In contrast, consider an equation in x and y such as

$$xy - 1 = 0.$$

The function $y = f(x)$ is said to be *implicitly* defined by this equation if the condition

$$x \cdot f(x) - 1 = 0$$

is satisfied. Sometimes an explicit expression for an implicitly defined function can be determined. For example, by solving $xy - 1 = 0$ for y we obtain

$$y = \frac{1}{x}.$$

Thus $f(x) = 1/x$ is an explicit expression for the function f defined by $xy - 1 = 0$.

Sometimes an equation in x and y defines more than one function. For example, if $x^2 + y^2 - 1 = 0$ is solved for y, we obtain

$$y = \sqrt{1 - x^2} \quad \text{and} \quad y = -\sqrt{1 - x^2}.$$

$f(x) = \sqrt{1 - x^2}$ and $g(x) = -\sqrt{1 - x^2}$ are both implicitly defined by $x^2 + y^2 - 1 = 0$.

Sometimes an equation implicitly defines y as a function of x and it is either not possible or very difficult to express y explicitly as a function of x.

Let us suppose that we are given an equation in x and y which implicitly defines y as a function of x and that this function is differentiable. Consider the problem of determining its derivative. One approach, of course, is to solve the equation for y and obtain an explicit expression for y which can be differentiated with the aid of the tools we have developed. This approach is not always possible and, when possible, is sometimes laborious. For the function $y = 1/x$ defined by $xy - 1 = 0$,

---

*This section may be omitted without disturbing continuity.

this approach yields

$$\frac{dy}{dx} = -\frac{1}{x^2}.$$

Another approach, called *implicit differentiation,* is to differentiate each term of the given equation in x and y, keeping in mind that y stands for a value of the implicitly defined function. The equation obtained can then be solved for the derivative of the implicitly defined function. The result will usually yield $\frac{dy}{dx}$ in terms of x and y. For the function $y = 1/x$, implicitly defined by $xy - 1 = 0$, implicit differentiation yields

$$\frac{d(xy)}{dx} - \frac{d(1)}{dx} = \frac{d(0)}{dx}.$$

By the product theorem we obtain

$$x\frac{d(y)}{dx} + y\frac{d(x)}{dx} - 0 = 0$$

$$x\frac{dy}{dx} + y = 0$$

$$x\frac{dy}{dx} = -y$$

$$\frac{dy}{dx} = -\frac{y}{x}. \tag{1}$$

Since we know that $y = 1/x$, we can substitute $1/x$ for y in equation (1) and express $\frac{dy}{dx}$ in terms of x. We obtain

$$\frac{dy}{dx} = \frac{-\dfrac{1}{x}}{x} = -\frac{1}{x^2}.$$

EXAMPLE 27. By implicit differentiation, determine the derivative of the function $y = f(x)$ defined implicitly by

$$x^2y - y - 7 = 0.$$

Solution. Term by term differentiation yields

$$\frac{d(x^2y)}{dx} - \frac{dy}{dx} - \frac{d(7)}{dx} = \frac{d(0)}{dx}.$$

By the product theorem we obtain

$$x^2 \frac{dy}{dx} + y \frac{d(x^2)}{dx} - \frac{dy}{dx} - 0 = 0$$

$$x^2 \frac{dy}{dx} + 2xy - \frac{dy}{dx} = 0.$$

Factoring $\dfrac{dy}{dx}$ yields

$$\frac{dy}{dx}(x^2 - 1) + 2xy = 0$$

$$\frac{dy}{dx}(x^2 - 1) = -2xy$$

$$\frac{dy}{dx} = \frac{-2xy}{x^2 - 1}.$$

EXAMPLE 28. $x^2 + y^2 - 1 = 0$ implicitly defines two differentiable functions, $y = \sqrt{1 - x^2}$ and $y = -\sqrt{1 - x^2}$. By implicit differentiation, determine the derivatives of these functions.

Solution. Term by term differentiation of $x^2 + y^2 - 1 = 0$ yields

$$\frac{d(x^2)}{dx} + \frac{d(y^2)}{dx} - \frac{d(1)}{dx} = \frac{d(0)}{dx}.$$

Since y is a function of x, we must use the chain rule to differentiate y^2.

$$2x + 2y \frac{dy}{dx} = 0$$

$$2y \frac{dy}{dx} = -2x$$

$$\frac{dy}{dx} = -\frac{x}{y}$$

Thus the derivative of $y = \sqrt{1 - x^2}$ is

$$\frac{dy}{dx} = \frac{-x}{\sqrt{1 - x^2}}$$

and the derivative of $y = -\sqrt{1 - x^2}$ is

$$\frac{dy}{dx} = \frac{x}{\sqrt{1 - x^2}}.$$

Exercises

By implicit differentiation, determine $\dfrac{dy}{dx}$ for the functions implicitly defined by the following equations.

145. $x^2y - 4 = 0$
146. $4x^3y + 8 = 0$
147. $x^2y - x + 14 = 0$
148. $4x^2y + 4x - 5 = 0$
149. $3x^3y + x^2y + 4 = 0$
150. $x^2 + y^2 - 16 = 0$
151. $4x^2 + 9y^2 - 25 = 0$
152. $xy + x^2 = 10$
153. $xy = 8$
154. $x^2 + y^2 = 25$
155. $x^2y^2 = 9$
156. $y^2 - 4x = 0$

17. Concepts which exhibit the derivative structure

Now that we have developed tools to help us determine derivatives, let us return to some of the concepts which exhibit the derivative structure and apply the results to their study. Concepts which exhibit the derivative structure include the following.

Instantaneous Velocity. If $d = f(t)$ is a time-distance function which describes the motion of some object, then $f'(t)$ is defined as the instantaneous velocity of the object at time t.

For $f(t) = 3t^3 + 4$, the instantaneous velocity at time t is $f'(t) = 9t^2$. The instantaneous velocity at time $t = 2$ is $f'(2) = 36$.

Instantaneous Acceleration. If $v = v(t)$ describes the instantaneous velocity at time t of an object in motion, then $v'(t)$ is defined as the instantaneous acceleration of the object at time t.

For $v(t) = 9t^2$, the instantaneous acceleration at time t is $v'(t) = 18t$. The instantaneous acceleration at time $t = 2$ is $v'(2) = 36$.

Tangent Line to a Curve. The tangent line to the graph of a function f at point $P(x, f(x))$ is defined as the line through P with slope $f'(x)$.

Thus the tangent line to the graph of $f(x) = 1/x$ at $P(x, 1/x)$ has slope $f'(x) = -(1/x^2)$. The tangent line to this curve at $P(1,1)$ has slope $f'(1) = -1$ (see Figure 17).

Marginal Revenue. If $R = R(x)$ is the total revenue function of a monopolist supplying x units of a certain commodity to a market per unit time, then $R'(x)$ is defined as the marginal revenue for an output of x units.

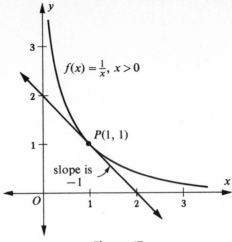

Figure 17

To illustrate, for the total revenue function $R(x) = 800x - 3x^2$, the marginal revenue for an output of x units is $R'(x) = 800 - 6x$. The marginal revenue for an output of 100 units is $R'(100) = 200$.

Marginal Cost. If $c = c(x)$ is the total cost function of a firm producing x units of a certain commodity per unit time, then $c'(x)$ is defined as the marginal cost for an output of x units.

For the total cost function $c(x) = \frac{1}{5}x^2 + 10x + 200$, the marginal cost for an output of x units is $c'(x) = \frac{2}{5}x + 10$. The marginal cost for an output of 50 units is $c'(50) = 30$.

Elasticity of Demand. Let $x = f(p)$ denote the demand function for a commodity in a certain market which expresses x, the quantity purchased per unit time, as a function of p, the unit price. If the price of the commodity changes from p_0 to p, then the percentage changes in quantity and price are

$$\frac{f(p) - f(p_0)}{f(p_0)} \times 100 \text{ percent} \quad \text{and} \quad \frac{p - p_0}{p_0} \times 100 \text{ percent}$$

respectively. The ratio

$$\frac{\dfrac{f(p) - f(p_0)}{f(p_0)}(100)}{\dfrac{p - p_0}{p_0}(100)} = \frac{p_0}{f(p_0)}\left(\frac{f(p) - f(p_0)}{p - p_0}\right)$$

expresses the ratio of the percentage change in quantity demanded to the given percentage change in price. If $f'(p_0)$ exists, then

$$\underset{p \to p_0}{\text{limit}} \frac{p_0}{f(p_0)} \left(\frac{f(p) - f(p_0)}{p - p_0} \right) = \frac{p_0}{f(p_0)} \underset{p \to p_0}{\text{limit}} \frac{f(p) - f(p_0)}{p - p_0} = \frac{p_0}{f(p_0)} f'(p_0).$$

The *point elasticity of demand at price* p_0, denoted by the Greek letter η ("eta"), is defined by

$$\eta = -\frac{p_0}{f(p_0)} f'(p_0).$$

The introduction of the negative sign is to make η a positive quantity. Its introduction does not affect the elasticity concept in any material way. Some economists, however, do not use it.

The point elasticity of demand at price p_0 serves as a measure of the responsiveness of demand for a commodity to a small change in price from p_0 to p. To illustrate, consider the demand function

$$f(p) = 120 - \tfrac{1}{5}p.$$

$f'(p_0) = -\tfrac{1}{5}$. Thus the point elasticity of demand for price p_0 is

$$\eta = -\frac{p_0}{f(p_0)} f'(p_0) = -\frac{p_0}{120 - \frac{1}{5}p_0} \left(-\frac{1}{5} \right) = \frac{p_0}{600 - p_0}.$$

The elasticity of demand for a market price of 50 monetary units per quantity unit is

$$\eta = \frac{50}{600 - 50} = \frac{1}{11}.$$

There are three elasticity situations to be distinguished.

1. If $\eta > 1$, demand is said to be *elastic*. A given percentage change in price brings with it a greater percentage change in quantity demanded.
2. If $\eta = 1$, demand is said to have *unit elasticity*. A given percentage change in price brings with it an equal percentage change in quantity demanded.
3. If $\eta < 1$, demand is said to be *inelastic*. A given percentage change in price brings with it a smaller percentage change in demand.

To illustrate, let us return to the demand function $f(p) = 120 - \tfrac{1}{5}p$, $0 < p < 600$ (see Figure 18). As we have noted, point elasticity of demand for price p is

$$\eta = \frac{p}{600 - p}.$$

For $\eta = 1$ we have

$$\frac{p}{600 - p} = 1, \quad p = 600 - p, \quad 2p = 600, \quad p = 300.$$

Thus $\eta = 1$ when $p = 300$. Also, $\eta > 1$ when $p > 300$, and $\eta < 1$ when $p < 300$.

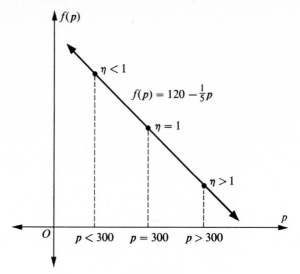

Figure 18

Demand, in this situation, is inelastic when $p < 300$, of unit elasticity when $p = 300$, and elastic when $p > 300$.*

Elasticity of Total Cost. If $c = c(x)$ is the total cost function of a firm producing x units of a certain commodity per unit time, then the elasticity of total cost for output x, denoted by the Greek letter κ ("kappa"), is defined by

$$\kappa = \frac{x}{c(x)} c'(x).$$

Thus for the total cost function $c(x) = x^2 + x + 200$, the elasticity of total cost for output x is

$$\kappa = \frac{x}{x^2 + x + 200}(2x + 1) = \frac{x(2x + 1)}{x^2 + x + 200}.$$

There are three situations to be distinguished.

1. If $\kappa < 1$ at a given output, then the situation is one of increasing returns in which a small increase in output is obtained from a less than proportionate increase in cost.
2. If $\kappa = 1$ at a given output, then the situation is one of constant returns. A small increase in output is obtained from a proportionate increase in cost.

---

*For further discussion of price elasticity of demand, see R. G. D. Allen, *Mathematical Analysis for Economists* (New York: Macmillan Company, 1939), Section 10.6; and works on price theory and microeconomic theory such as D. S. Watson, *Price Theory and Its Uses,* 2d ed. (Boston: Houghton Mifflin Co., 1968), especially p. 137; C. E. Ferguson, *Microeconomic Theory* (Homewood, Illinois: Richard D. Irwin, Inc., 1969); J. M. Henderson and R. E. Quandt, *Microeconomic Theory: A Mathematical Approach* (New York: McGraw-Hill, 1958).

3. If $\kappa > 1$ at a given output, then the situation is one of decreasing returns in which a small increase in output is obtained from a proportionately greater increase in cost.*

Marginal Propensity to Consume. In macroeconomic theory, if $C = C(Y)$ expresses consumption C as a function of income Y, then the marginal propensity to consume for an income Y is defined as $c = \dfrac{dC}{dY}$.†

EXAMPLE 29. A coffee producer turns out an output of x tons a day at a total cost of

$$c(x) = \frac{x^2 + 300x}{x + 200}$$

hundred dollars. Determine the marginal cost function, the marginal cost of an output of 10 tons a day, and the elasticity of total cost for an output of 10 tons a day.

Solution. The marginal cost function is $c'(x)$. By the quotient theorem we have

$$c'(x) = \frac{(x + 200)\dfrac{d(x^2 + 300x)}{dx} - (x^2 + 300x)\dfrac{d(x + 200)}{dx}}{(x + 200)^2}.$$

Since

$$\frac{d(x^2 + 300x)}{dx} = 2x + 300 \quad \text{and} \quad \frac{d(x + 200)}{dx} = 1$$

we obtain

$$c'(x) = \frac{(x + 200)(2x + 300) - (x^2 + 300x)(1)}{(x + 200)^2}.$$

Multiplying and combining terms yields the marginal cost function

$$c'(x) = \frac{x^2 + 400x + 60,000}{(x + 200)^2}.$$

The marginal cost for an output of 10 tons a day is $c'(10) = 1.45$. Elasticity of total cost is

$$\kappa = \frac{x}{c(x)} c'(x).$$

Thus the elasticity of total cost for an output of 10 tons of coffee a day is

$$\kappa = \frac{10}{c(10)} c'(10).$$

---

*For further discussion of elasticity of total cost, see Allen, *Mathematical Analysis for Economists,* Section 10.8.
†For further discussion, see R. G. D. Allen, *Macro-Economic Theory: A Mathematical Treatment* (New York: St. Martin's Press, 1967), p. 19.

$c(10) = 14.76$ and $c'(10) = 1.45$; therefore we obtain

$$\kappa = \frac{10}{14.76}(1.45) = 0.98$$

for the elasticity of total cost.

EXAMPLE 30. For a certain community it is found that x, the number of gallons of orange juice purchased a day in the summer months, depends on p, the price in cents per gallon, according to

$$x = (90,000)e^{-(p/30)}.$$

Determine the elasticity of demand for a price of 30¢ a gallon.

Solution. Elasticity of demand is

$$\eta = -\frac{p}{x} \cdot \frac{dx}{dp}.$$

Now

$$\frac{dx}{dp} = (90,000)\frac{d(e^{-(p/30)})}{dp} = (90,000)\frac{dy}{dp} \qquad (1)$$

where $y = e^{-(p/30)}$. To determine $\frac{dy}{dp}$ we must use the chain rule. If we let $u = -(p/30)$, then our second auxiliary function is $y = e^u$. Since $y = e^u$,

$$\frac{dy}{du} = e^u.$$

Since $u = -(p/30)$,

$$\frac{du}{dp} = -\frac{1}{30}.$$

Thus by the chain rule we have

$$\frac{dy}{dp} = \frac{dy}{du} \cdot \frac{du}{dp} = e^u\left(-\frac{1}{30}\right).$$

Replacing u by $-(p/30)$ yields

$$\frac{dy}{dp} = \left(-\frac{1}{30}\right)e^{-(p/30)}.$$

By substituting this result into equation (1) we obtain

$$\frac{dx}{dp} = (90,000)\left(-\frac{1}{30}\right)e^{-(p/30)} = -3000\,e^{-(p/30)}.$$

For $p = 30$, $x = (90,000)e^{-1}$ and $\frac{dx}{dp} = (-3,000)e^{-1}$. Thus the elasticity of demand for a price of 30¢ a gallon is

$$\eta = \frac{-30(-3000)e^{-1}}{(90{,}000)e^{-1}} = 1.$$

EXAMPLE 31. The inverse of the orange-juice demand function

$$x = (90{,}000)e^{-(p/30)}$$

discussed in the previous example is

$$p = 30 \ln \frac{90{,}000}{x}.$$

The total revenue function for the orange-juice producer is obtained by multiplying the price of a unit p by output x. Thus we have

$$R = xp = 30x \ln \frac{90{,}000}{x}.$$

Determine the marginal revenue function for this orange-juice producer.

Solution. The marginal revenue function is $R'(x)$. By the product theorem we have

$$R'(x) = 30x \frac{d\left(\ln \frac{90{,}000}{x}\right)}{dx} + \ln\left(\frac{90{,}000}{x}\right)\frac{d(30x)}{dx}. \tag{1}$$

Now $\dfrac{d(30x)}{dx} = 30$. By using the chain rule we obtain

$$\frac{d\left(\ln \frac{90{,}000}{x}\right)}{dx} = -\frac{1}{x}.$$

Substituting these results into equation (1) yields

$$R'(x) = -30 + 30 \ln \frac{90{,}000}{x}$$

for the marginal revenue function.

EXAMPLE 32. One of the cost-function types studied in economics is

$$c(x) = \sqrt{Ax + B} + D$$

where A, B, and D are positive constants. Determine the marginal cost function for this cost-function type.

Solution. By definition marginal cost is $c'(x)$. We have

$$c(x) = (Ax + B)^{\frac{1}{2}} + D.$$

By the sum theorem,

$$c'(x) = \frac{d(Ax + B)^{\frac{1}{2}}}{dx} + \frac{d(D)}{dx}. \tag{1}$$

Now $\dfrac{d(D)}{dx} = 0$. From the chain rule we obtain

$$\frac{d(Ax + B)^{\frac{1}{2}}}{dx} = \frac{A}{2\sqrt{Ax + B}}.$$

Substituting these results into equation (1) yields the marginal cost function

$$c'(x) = \frac{A}{2\sqrt{Ax + B}}.$$

A relation between demand, total revenue, and marginal revenue*

An important relation in economics, which we will establish here, is

$$\frac{dR}{dx} = p\left(1 - \frac{1}{\eta}\right)$$

where $p = g(x)$ and its inverse $x = f(p)$ are demand functions which relate p, the unit price, and x, the quantity bought by the market per unit time. $R = xp$ is the total revenue function for the firm producing output x, $\dfrac{dR}{dx}$ is marginal revenue, and $\eta = -\dfrac{p}{x} \cdot \dfrac{dx}{dp}$ is elasticity of demand.

Proof. Since $R = xp$, where $p = g(x)$, we have by the product theorem

$$\frac{dR}{dx} = x\frac{dp}{dx} + p\frac{dx}{dx} = x\frac{dp}{dx} + p = p + x\frac{dp}{dx}. \tag{1}$$

By factoring p from the right member of equation (1) we obtain

$$\frac{dR}{dx} = p\left(1 + \frac{x}{p} \cdot \frac{dp}{dx}\right). \tag{2}$$

To complete our proof we must show that

$$\frac{x}{p} \cdot \frac{dp}{dx} = -\frac{1}{\eta}.$$

Since

$$\eta = -\frac{p}{x} \cdot \frac{dx}{dp}, \quad \text{then} \quad -\frac{1}{\eta} = \frac{x}{p} \cdot \frac{1}{\dfrac{dx}{dp}}.$$

---

*This topic may be omitted without disturbing continuity.

But

$$\frac{1}{\dfrac{dx}{dp}} = \frac{dp}{dx}$$

since $x = f(p)$ and $p = g(x)$ are inverses. Thus we obtain

$$-\frac{1}{\eta} = \frac{x}{p} \cdot \frac{dp}{dx}.$$

Substituting this result into equation (2) yields

$$\frac{dR}{dx} = p\left(1 - \frac{1}{\eta}\right).^*$$

Exercises

157. Describe three concepts which exhibit the derivative structure other than the ones discussed in this section.

158. A steel plant produces x tons of steel a day at a total cost of $c(x) = \frac{1}{12}x^3 - x^2 + 25x + 800$ dollars. Determine the marginal cost function, the marginal cost for an output of 12 tons a day, and the elasticity of total cost for an output of 12 tons a day.

159. Find the tangent line to the graph of $f(x) = 1/(x^2 + 1)$ at the point $P(-1,\frac{1}{2})$.

160. The demand for chocolate in a certain city is described by the demand functions

$$x = \frac{800}{p + 4} - 10 \quad \text{and} \quad p = \frac{800}{x + 10} - 4$$

where x represents the market demand for chocolate in thousands of pounds per day and p is the price of chocolate in dollars per thousand-pounds. Determine the elasticity of demand for a price of \$36 per thousand-pounds. Assuming that the chocolate is produced by a monopolist, determine his total revenue and marginal revenue functions. What is the marginal revenue corresponding to a price of \$36 per thousand-pounds?

161. The motion of a certain object is described by the time-distance function $d = 3t^3 + 5t$. Find the instantaneous velocity and instantaneous acceleration of the object at time t.

162. A cost-function type studied in economics is $c(x) = Ax^2 + Bx + C$, where A, B, and C are positive constants. Determine the marginal cost for this cost-function type and show that elasticity of total cost is given by

$$\kappa = \frac{2Ax^2 + Bx}{Ax^2 + Bx + C}.$$

---

*For further discussion of this relation, see Allen, *Mathematical Analysis for Economists;* Ferguson, *Microeconomic Theory,* Section 9.2; and Watson, *Price Theory and Its Uses,* Part Five.

163. A tea plantation produces x tons of tea a day at a cost of

$$c(x) = \frac{x^2(x + 50)}{50(x + 200)} + 75$$

hundred dollars. Determine the marginal cost function and the marginal cost for an output of 10 tons of tea a day.

164. The demand for milk in a certain city is described by the demand functions

$$x = \frac{50 - \sqrt{p}}{2} \quad \text{and} \quad p = (50 - 2x)^2$$

where x represents the average market demand for milk per household in gallons per month and p is the price of milk in cents per gallon. Determine the elasticity of demand for a price of 64¢ per gallon. Assuming that the milk supply is controlled by a monopolist, determine his total revenue and the marginal revenue function. What is the marginal revenue corresponding to a price of 64¢ per gallon?

165. Find the tangent line to the graph of $f(x) = 3x/(2x + 1)$ at the point $P(1,1)$.

18. Equivalent formulations of the derivative concept*

We defined the derivative of function f at value a as

$$f'(a) = \lim_{x \to a} \frac{f(x) - f(a)}{x - a}.$$

If we let $h = x - a$, then $x = a + h$, and we can write

$$f'(a) = \lim_{h \to 0} \frac{f(a + h) - f(a)}{h}.$$

In this formulation of the derivative concept, $f'(x)$ is expressed by

$$f'(x) = \lim_{h \to 0} \frac{f(x + h) - f(x)}{h}.$$

If instead of h the symbol Δx (read "delta x") is used, we obtain

$$f'(x) = \lim_{\Delta x \to 0} \frac{f(x + \Delta x) - f(x)}{\Delta x}.$$

h, or Δx, represents the *change in* x. The corresponding *change in the function f,* denoted by Δf, is

$$\Delta f = f(x + \Delta x) - f(x).$$

---

*This section may be omitted without disturbing continuity.

In terms of this symbolism we have

$$f'(x) = \lim_{\Delta x \to 0} \frac{\Delta f}{\Delta x}.$$

When y is used to denote $f(x)$, then the corresponding change in the function is denoted by Δy, and we have

$$\Delta y = f(x + \Delta x) - f(x).$$

Thus

$$f'(x) = \lim_{\Delta x \to 0} \frac{\Delta y}{\Delta x}.$$

EXAMPLE 33. For $y = x^2$, find $\dfrac{dy}{dx}$ by using the Δx, Δy notation for the derivative.

Solution.

$$f(x) = x^2$$
$$f(x + \Delta x) = (x + \Delta x)^2 = x^2 + 2x(\Delta x) + (\Delta x)^2$$
$$\Delta y = f(x + \Delta x) - f(x) = 2x(\Delta x) + (\Delta x)^2$$
$$\frac{\Delta y}{\Delta x} = \frac{2x(\Delta x) + (\Delta x)^2}{\Delta x} = \frac{(\Delta x)(2x + \Delta x)}{\Delta x} = 2x + \Delta x, \quad \Delta x \neq 0$$

Thus

$$\frac{dy}{dx} = \lim_{\Delta x \to 0} \frac{\Delta y}{\Delta x} = \lim_{\Delta x \to 0} (2x + \Delta x) = 2x.$$

19. Higher derivatives

The derivative f' of a function f is a function. Thus it too may have a derivative. The derivative of f' is denoted by f'', $D^2 f$, $\dfrac{d^2 f}{dx^2}$, or $\dfrac{d^2 y}{dx^2}$, and is called the *second derivative of f*. The derivative of f'' is called the *third derivative of f* and is denoted by f''', $D^3 f$, $\dfrac{d^3 f}{dx^3}$, or $\dfrac{d^3 y}{dx^3}$. To find the second or third derivative, we use the theorems and rules established for a derivative, with f' and f'' as the functions.

EXAMPLE 34. Find the second derivative of $f(x) = x^4$, $f(x) = 1/x$, and $f(x) = 2x^2 + 3x - 4$.

Solution.

$$f(x) = x^4 \qquad\qquad f'(x) = 4x^3 \qquad\quad f''(x) = 12x^2$$

$$f(x) = \frac{1}{x} = x^{-1} \qquad\quad f'(x) = -x^{-2} \qquad\quad f''(x) = 2x^{-3} = \frac{2}{x^3}$$

$$f(x) = 2x^2 + 3x - 4 \qquad f'(x) = 4x + 3 \qquad\quad f''(x) = 4$$

Exercises

Find the second derivative of each of the following functions at the specified value.

166. $f(x) = x^3 + 2x + 1, \quad x = 2$
167. $f(x) = 3x^2 + 4x - 7, \quad x = 1$
168. $f(x) = \ln x, \quad x = 10$
169. $f(x) = 3\, e^x, \quad x = 2$
170. $\bar{c}(x) = \dfrac{1}{4}x + \dfrac{3}{2} + \dfrac{1}{x}, \quad x = 2$
171. $P(x) = -\frac{21}{4}x^2 + \frac{1197}{2}x - 1, \quad x = 57$
172. $\bar{c}(x) = \dfrac{1}{10}x + 2 + \dfrac{10}{x}, \quad x = 10$
173. $P(x) = -x^2 + 900x - 10, \quad x = 450$
174. $\bar{c}(x) = \dfrac{1}{100}x + 3 + \dfrac{100}{x}, \quad x = 100$
175. $P(x) = -x^2 + 1002x - 100, \quad x = 501$
176. $P(x) = -\frac{1}{3}x^3 + 7x^2 + 1800x - 100, \quad x = 50$
177. $P(x) = -\frac{1}{3}x^3 + 8x^2 + 960x - 150, \quad x = 40$

3

Optimization problems

20. Extreme values of functions

There are many situations in which one is led to seek the greatest or least value of a function. To illustrate, consider a chocolate producer for which

$$c(x) = x^2 + 2x + 100$$

is the total dollar cost of producing x tons of chocolate a day. The average cost function, obtained by dividing cost $c(x)$ by output x, is

$$\overline{c}(x) = x + 2 + \frac{100}{x}.$$

For what output is average cost $\overline{c}(x)$ a minimum? If the profit function of this firm is

$$P(x) = -101x^2 + 1198x - 100$$

for what output is profit $P(x)$ a maximum?

In this section, with the aid of the calculus, we will develop methods for answering questions of this sort. We begin by introducing some basic definitions.

A function f is said to have an *absolute maximum value* at c if $f(c) \geq f(x)$ for all values of x in the domain of definition of f. (See Figure 1.) $f(c)$ itself is said to be the *maximum value* of f.

Function f is said to have a *local maximum value* at c if there is an interval I containing c such that $f(c) \geq f(x)$ for any value of x in interval I which is also in the domain of definition of f. $f(c)$ is said to be a *local maximum value* of f.

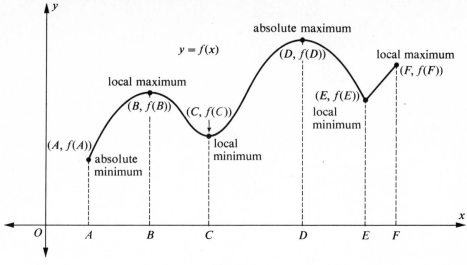

Figure 1

The concepts of *absolute minimum value* and *local minimum value* are defined in an analogous way.

The function f shown in Figure 1 has an absolute minimum value at A, an absolute maximum value at D, local minimum values at C and E, and local maximum values at B and F.

The maximum and minimum values of a function are called its *extreme values*. When does a function have extreme values? One answer can be given in terms of the concept of continuity. It can be shown that if a function is continuous at every point of an interval I, including the end points of I, then the function has an absolute maximum value and an absolute minimum value in I. If a function is not continuous at every point in an interval, then it may or may not have extreme values.

Where in its domain of definition does a function assume its extreme values? To obtain some insight into where extreme values occur, consider the functions whose graphs are shown in Figures 2 through 5. The extreme values of $f(x) = x + 1$, $0 \leq x \leq 2$ (Figure 2) occur at 0 and 2; the extreme value of $f(x) = x^2 + 1$ (Figure 3) occurs at 0, as is also the case with $f(x) = -x^2 + 1$ (Figure 4) and $f(x) = \begin{cases} x, & x \geq 0 \\ -x, & x < 0 \end{cases}$ (Figure 5).

What properties do these numbers have? 0 and 2 are end points of the domain of definition of $f(x) = x + 1$, $0 \leq x \leq 2$. 0 in the domain of definition of $f(x) = x^2 + 1$ yields the point $(0, 1)$ at which the tangent line to the graph of $f(x) = x^2 + 1$ is horizontal, which means that the derivative of $f(x) = x^2 + 1$ is zero at this value. A similar situation is the case for 0 in the domain of definition of $f(x) = -x^2 + 1$. On the other hand, the derivative of $f(x) = \begin{cases} x, & x \geq 0 \\ -x, & x < 0 \end{cases}$ is not defined at 0 (see Example 10, Section 9).

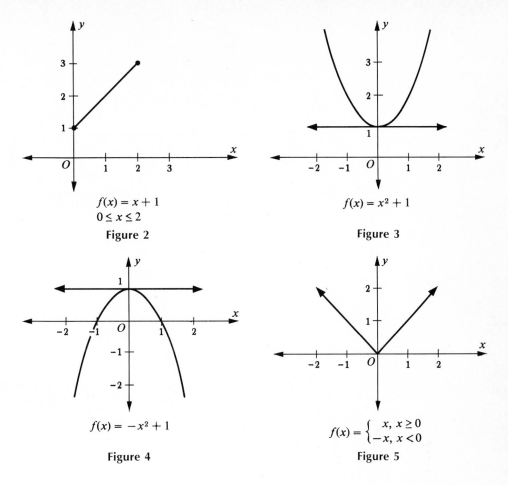

$f(x) = x + 1$
$0 \le x \le 2$

Figure 2

$f(x) = x^2 + 1$

Figure 3

$f(x) = -x^2 + 1$

Figure 4

$f(x) = \begin{cases} x, & x \ge 0 \\ -x, & x < 0 \end{cases}$

Figure 5

These examples show that extreme values can occur at numbers of the following types in the domain of f.

1. A number c such that $f'(c) = 0$.
2. A number c such that $f'(c)$ does not exist.
3. A number c which is an end point of the domain of definition of function f.

It can be shown that a function can have an extreme value only at numbers of these three types. We will call such numbers *critical values* of the function. Extreme values occur only at critical values. Thus the search for extreme values begins with a search for critical values. However, while every extreme value occurs at a critical value, not every critical value need necessarily yield an extreme value.

As an illustration, consider $f(x) = x^3 + 1$, whose graph is shown in Figure 6. 0 is a critical value (type 1) since $f'(0) = 0$. $f(0) = 1$ but 1 is not a maximum value since it is not true that $f(0) \ge f(x)$ for x in some interval containing 0; $f(x) > 1$ when x is positive. $f(0)$ is not a minimum value since it is not true that $f(0) \le f(x)$

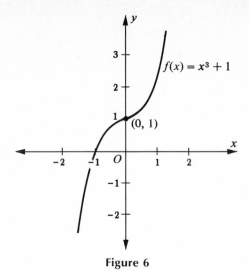

Figure 6

for x in some interval containing 0; $f(x) < 1$ when x is negative. Thus, while 0 is a critical value for $f(x) = x^3 + 1$, $f(0) = 1$ is not an extreme value.

Therefore to find the extreme values of a function, we must first find its critical values and then distinguish those which yield extreme values from those which do not. The real problem is with critical values of type 1 (c such that $f'(c) = 0$).

In fact, a method for distinguishing critical values of type 1 which give rise to extreme values from those which do not is suggested by two situations that we have already examined, $f(x) = x^2 + 1$ and $f(x) = -x^2 + 1$ (shown in Figures 3 and 4). 0 is a critical value for both of these functions. $f(0)$ is a minimum value of $f(x) = x^2 + 1$ and $f(0)$ is a maximum value of $f(x) = -x^2 + 1$. In both cases $f'(0) = 0$. But how does $f'(x)$ behave near the critical value 0?

Let us observe that in the case of $f(x) = x^2 + 1$, with a minimum value at 0, the derivative $f'(x) = 2x$. Thus the slope of the tangent line is negative for x less than the critical value 0 and positive for x greater than the critical number 0. That is, $f'(x) = 2x$ goes from negative to positive as x goes from values less than the critical value 0 to values greater than the critical value 0 (see Figure 7).

On the other hand, for $f(x) = -x^2 + 1$, just the opposite happens. The derivative $f'(x) = -2x$, and thus the slope of the tangent line is positive for x less than the critical value 0 and negative for x greater than the critical value 0. That is, $f'(x) = -2x$ goes from positive to negative as x goes from values less than the critical value 0 to values greater than the critical value 0 (see Figure 8).

These examples suggest the following first derivative test, which is indeed valid.

First Derivative Test. Let c denote a number with the property that $f'(c) = 0$. If there is an interval I containing c such that $f'(x)$ goes from negative to positive as x goes from values less than c in I to values greater than c in I, then $f(c)$ is a local minimum value of f. If $f'(x)$ goes from positive to negative, then $f(c)$ is a local

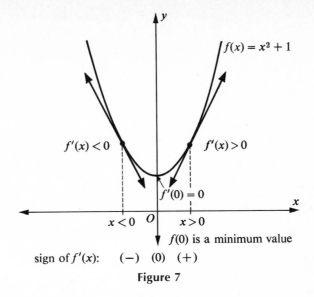

sign of $f'(x)$: $(-)$ (0) $(+)$

Figure 7

maximum value of f. If $f'(x)$ does not change sign, then $f(c)$ is not an extreme value of f.

In summary we have

$f(c)$ is a local maximum value if $f'(c) = 0$ and $f'(x)$ goes from $+$ to $-$;

$f(c)$ is a local minimum value if $f'(c) = 0$ and $f'(x)$ goes from $-$ to $+$;

$f(c)$ is not an extreme value when $f'(c) = 0$ but $f'(x)$ does not change sign.

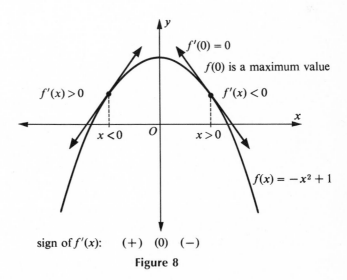

sign of $f'(x)$: $(+)$ (0) $(-)$

Figure 8

EXAMPLE 1. The average cost function for a chocolate producer is

$$\bar{c}(x) = x + 2 + \frac{100}{x}$$

dollars for an output of x tons of chocolate per day. Determine the output for which his average cost is a minimum.

Solution.

The domain of definition of $\bar{c}(x)$ is $x > 0$, subject to plant limitation. There are no end point critical values and there are no values of $x > 0$ at which $\bar{c}'(x)$ does not exist. To determine critical values of type 1 (values at which $\bar{c}'(x) = 0$), we must determine $\bar{c}'(x)$, set it equal to zero, and solve for x.

$$\bar{c}'(x) = 1 - \frac{100}{x^2}$$

Setting $\bar{c}'(x)$ equal to zero and solving for x yields

$$1 - \frac{100}{x^2} = 0$$
$$x^2 = 100$$
$$x = 10.$$

Thus $x = 10$ is a critical value of type 1.

To determine if $\bar{c}(10)$ is a minimum value, we must examine the behavior of

$$\bar{c}'(x) = 1 - \frac{100}{x^2} = \frac{x^2 - 100}{x^2}$$

for $x < 10$ and $x > 10$.

When $x < 10$, $x^2 < 100$ and $\bar{c}'(x) < 0$.
When $x > 10$, $x^2 > 100$ and $\bar{c}'(x) > 0$.

Since $\bar{c}'(10) = 0$ and $\bar{c}'(x)$ goes from negative to positive, it follows by the first derivative test that $\bar{c}(10)$ is a minimum value. Thus average cost is minimized when 10 tons of chocolate are produced daily.

EXAMPLE 2. The chocolate producer's profit function is

$$P(x) = -101x^2 + 1198x - 100$$

dollars for an output of x tons of chocolate a day. Determine the output for which profit is maximized.

Solution.

The domain of definition of $P(x)$ is $x > 0$, subject to plant limitation. There are no end point critical values and there are no values of $x > 0$ at which $P'(x)$ does not exist. To find critical values of type 1, we must determine $P'(x)$, set it equal to zero, and solve for x.

$$P'(x) = -202x + 1198$$

Setting $P'(x)$ equal to zero and solving for x yields

$$-202x + 1198 = 0$$

$$x = \frac{-1198}{-202} = 5.93.$$

Thus $x = 5.93$ is a critical value of type 1.

To determine if $P(5.93)$ is a maximum value, we must examine the behavior of

$$P'(x) = -202x + 1198 = -202(x - 5.93)$$

for $x < 5.93$ and $x > 5.93$. When $x < 5.93$,

$$P'(x) = \underbrace{-202}_{\text{neg.}}\underbrace{(x - 5.93)}_{\text{neg.}}.$$

$P'(x)$ is the product of two negative numbers and thus is positive. When $x > 5.93$,

$$P'(x) = \underbrace{-202}_{\text{neg.}}\underbrace{(x - 5.93)}_{\text{pos.}}.$$

$P'(x)$ is the product of a negative number and a positive number and thus is negative. Since $P'(5.93) = 0$ and $P'(x)$ goes from positive to negative, it follows by the first derivative test that $P(5.93)$ is a maximum value. Thus profit is maximized when 5.93 tons of chocolate are produced daily. The maximum daily profit is $P(5.93) = \$3453$.

Exercises

Use the first derivative test to determine the extreme values of the following functions.

1. $f(x) = 2x^2 + 3x + 6$
2. $f(x) = -3x^2 + 12x + 5$
3. $f(x) = -4x^2 + 24x + 3$
4. $f(x) = \frac{3}{2}x^2 + 9x + 8$
5. $f(x) = 3x^4 - 4x^3$
6. $f(x) = 2x^3 + 3x^2$
7. $f(x) = 2x^3 - 3x^2$
8. $f(x) = x^3 + \frac{7}{2}x^2 - 6x - 10$
9. $f(x) = \frac{2}{3}x^3 - \frac{9}{2}x^2 + 4x - 12$
10. $f(x) = \frac{1}{3}x^3 - \frac{1}{2}x^2 - 6x - 4$
11. $f(x) = e^{-x^2}$
12. $f(x) = \frac{1}{\sqrt{2\pi}} e^{-\frac{1}{2}x^2}$

Another test for extreme values which is sometimes easier to apply than the first derivative test is the second derivative test. The second derivative test is based on the fact that the concavity of a function is indicated by its second derivative.

The graph of the function shown in Figure 9 is said to be *concave upward* at point P since the graph is above its tangent line at P. The graph of the function shown in Figure 10 is below its tangent line at P and is thus said to be *concave downward* at P. The concavity of a function f is tied to the behavior of its second derivative in the following way. If $f''(x) > 0$ at each point of an interval, then the graph of f is concave upward at each point of the interval. If $f''(x) < 0$ at each point of an interval, then the graph of f is concave downward at each point of the interval.

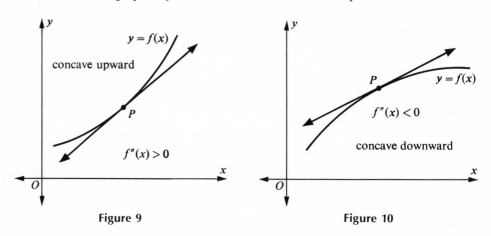

Figure 9   **Figure 10**

This information on the concavity of a function and the behavior of the second derivative of the function, together with what we know about the role of critical values in determining extreme values, lead us to the following second derivative test.

Second Derivative Test. Let c denote a number with the property that $f'(c) = 0$. If there is an interval I containing c such that $f'(x)$ exists in I, then

1. $f(c)$ is a *local minimum value* for f if $f''(c) > 0$;
2. $f(c)$ is a *local maximum value* for f if $f''(c) < 0$.

If $f''(c) = 0$, then no conclusion can be drawn.

EXAMPLE 3. Consider again the chocolate firm's average cost function

$$\bar{c}(x) = x + 2 + \frac{100}{x}, \quad x > 0.$$

In Example 1 we saw that

$$\bar{c}'(x) = 1 - \frac{100}{x^2}$$

and $\bar{c}'(10) = 0$. That is, $x = 10$ is a critical value of type 1. To determine if $x = 10$ yields an extreme value by the second derivative test, we must determine $\bar{c}''(10)$.

$$\bar{c}''(x) = \frac{200}{x^3}$$

Thus $\bar{c}''(10) > 0$, and it follows from the second derivative test that $\bar{c}(10)$ is a minimum value.

E X A M P L E   4.   Consider again the chocolate firm's profit function

$$P(x) = -101x^2 + 1198x - 100.$$

In Example 2 we saw that

$$P'(x) = -202x + 1198$$

and that $P'(5.93) = 0$. To determine if $x = 5.93$ yields an extreme value by the second derivative test, we must determine $P''(5.93)$.

$$P''(x) = -202$$

for all values of x. Thus $P''(5.93) = -202$, so $P''(5.93) < 0$, and it follows from the second derivative test that $P(5.93)$ is a maximum value.

If the second derivative of a function is difficult to determine, then practical considerations may favor the first derivative test. If $f'(c) = 0$ and $f''(c) = 0$, then the second derivative test is inconclusive and the first derivative test must be used.

To illustrate the inconclusiveness of the second derivative test when $f''(c) = 0$, consider $f(x) = x^3 + 1$, $f(x) = x^4 + 1$, and $f(x) = -x^4 + 1$. (See Figures 11, 12, and 13.) 0 is a critical value of type 1 for each of these functions. In each case

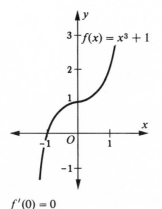

$f'(0) = 0$
$f''(0) = 0$
$f(0)$ is not an extreme value

Figure 11

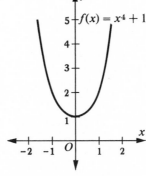

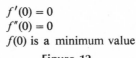

$f'(0) = 0$
$f''(0) = 0$
$f(0)$ is a minimum value

Figure 12

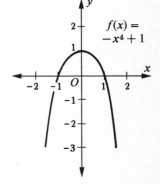

$f'(0) = 0$
$f''(0) = 0$
$f(0)$ is a maximum value

Figure 13

$f''(0) = 0$. As we have seen, $f(0)$ is not an extreme value for $f(x) = x^3 + 1$. $f(0)$ is a minimum value for $f(x) = x^4 + 1$ and a maximum value for $f(x) = -x^4 + 1$.

Exercises

For each of the following average cost functions, determine the output for which average cost is minimized.

13. $\bar{c}(x) = \dfrac{1}{100}x + 3 + \dfrac{100}{x}$

14. $\bar{c}(x) = \dfrac{1}{10}x + 2 + \dfrac{10}{x}$

15. $\bar{c}(x) = \dfrac{1}{4}x + \dfrac{3}{2} + \dfrac{1}{x}$

For each of the following profit functions, determine the output for which profit is maximized.

16. $P(x) = -x^2 + 900x - 10$

17. $P(x) = -x^2 + 1002x - 100$

18. $P(x) = -\frac{1}{3}x^3 + 7x^2 + 1800x - 100$

19. $P(x) = -\frac{1}{3}x^3 + 8x^2 + 960x - 150$

20. $P(x) = -\frac{21}{4}x^2 + \frac{1197}{2}x - 1$

For each of the following total revenue functions, determine the output for which total revenue is maximized.

21. $R(x) = 1920x - x^2$

22. $R(x) = 1060x - x^2$

23. $R(x) = 1020x - 3x^2$

24. A manufacturer is planning to market drums of oil. Each drum is to have a volume of 12 cubic feet, and the material needed to make the drums costs 10¢ per square foot. Determine the drum dimensions for which cost of material is minimized (see Exercise 22, Section 1).

25. A lamp manufacturer has fixed expenses of $1000 per week. His production cost is $5 per lamp. It is estimated that if the selling price of a lamp is x dollars, then $450 - 10x$ lamps would be sold per week. Determine the selling price that would maximize weekly income, the maximum weekly income, and the optimum number of lamps made and sold per week (see Exercise 21, Section 1).

26. The Ramune Savings Bank is trying to determine the interest rate that it should pay on money deposited with the bank in a certain time period. It is assumed that the amount of money that people will deposit is proportional to the interest paid by the bank (that is, that the amount A of additional money deposited is equal to kx, where k is a constant and x is the interest rate). The bank further assumes that it can earn 15% on the money that it raises. With respect to these

assumptions, determine the interest rate that it should offer so as to maximize its profit.

27. A nursery school wishes to construct a rectangular play region with perimeter 1000 feet. What are the dimensions of the play region with the largest area?

28. The Rasa Bus Company uses 40 buses to service a certain route. Each bus brings in an average of $1000 per day in fares. Studies indicate that if additional buses are assigned to the route, the amount in fares brought in by each bus will drop by $20 for each additional bus that is added. If this is the case, how many additional buses should be assigned to the route so that the total daily amount in fares obtained by the bus company is maximized?

29. A food chain has 50 stores in New York City, each doing an average of $30,000 worth of business per day. A study made by a consulting firm concluded that if new stores are opened in locations available to the chain, then the average amount of business done by each store will drop by $500 for each new store that is opened. Assuming this to be the case, how many new stores should be opened so that the total daily amount of business done by the chain is maximized?

30. The design of the Vroman Institute, at which studies of blood proteins and blood diseases are to be undertaken, calls for a rectangular structure with 20,000 square feet of floor space. The building material to be used for the front of the building is, per linear foot, three times as expensive as the material to be used for the other three sides. What should the dimensions of the building be for the material cost of the four walls to be a minimum?

31. The Andrius Publishing Company is planning to print 20,000 copies of a brochure to advertise a calculus book that they are publishing. The type plate containing the printed matter is to be 48 square inches, the top and bottom margins are each to be $1\frac{1}{2}$ inches, and the side margins are each to be 2 inches. Because of a paper shortage the company would like to use the least amount of paper possible. What should the dimensions of the type plate be so that the area of the brochure page is minimized?

32. A pipe line to transport oil is to connect cites A and B (shown in Figure 14) which are located on opposite sides of a river 5 miles wide. Site A is located 20 miles downstream from point C which is directly across the river from B.

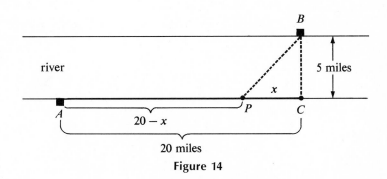

Figure 14

Constructing the pipe line along the line from A to C on the bank of the river would cost \$4000 per mile, while laying pipe across the river costs \$5000 per mile. What is the most economical route that can be used in constructing the pipe line? That is, at what point P on the bank should the pipe be laid across the river to B after having been laid along the river bank from A to P if the total cost is to be minimized? What is the minimum cost?

21. Profit-maximization behavior of a firm

Consider a monopolist producing a certain commodity for which

$$c = c(x)$$

is the total cost function for the production of x units (per unit time), and

$$R = R(x)$$

is the total revenue function for the production and sale of x units. A fundamental proposition of economics, which we will prove here, is that the monopolist's profit is maximized at that output for which

$$\frac{dR}{dx} = \frac{dc}{dx}$$

that is, that output for which marginal revenue is equal to marginal cost.

Proof. The profit function $P(x)$ of the monopolist is the difference between his total revenue and total cost functions.

$$P(x) = R(x) - c(x)$$

The domain of definition of $P(x)$ is assumed to be $x > 0$, subject to plant limitation. There are no end point critical values and, since $\frac{dR}{dx}$ and $\frac{dc}{dx}$ exist, $\frac{dP}{dx}$ exists for all x under consideration. Thus to find the value of x which maximizes profit, we determine $\frac{dP}{dx}$, let $\frac{dP}{dx} = 0$, and solve for x. From

$$P(x) = R(x) - c(x)$$

and the difference theorem for derivatives, we obtain

$$\frac{dP}{dx} = \frac{dR}{dx} - \frac{dc}{dx}.$$

Setting $\frac{dP}{dx}$ equal to zero yields

$$\frac{dR}{dx} - \frac{dc}{dx} = 0$$

and it follows that

$$\frac{dR}{dx} = \frac{dc}{dx}.$$

Thus the value of x for which $\frac{dP}{dx} = 0$ is the value for which marginal revenue equals marginal cost. Now

$$\frac{d^2P}{dx^2} = \frac{d^2R}{dx^2} - \frac{d^2c}{dx^2}.$$

By the second derivative test, if

$$\frac{d^2P}{dx^2} = \frac{d^2R}{dx^2} - \frac{d^2c}{dx^2}$$

and

$$\frac{d^2R}{dx^2} - \frac{d^2c}{dx^2} < 0$$

or equivalently

$$\frac{d^2R}{dx^2} < \frac{d^2c}{dx^2}$$

at x, then x yields a maximum value for the profit function $P(x) = R(x) - c(x)$. Thus profit is maximized at that output x at which marginal revenue is equal to marginal cost provided that

$$\frac{d^2R}{dx^2} < \frac{d^2c}{dx^2}.$$

EXAMPLE 5. Let us recall that the total cost function of our chocolate-producing monopolist is

$$c(x) = x^2 + 2x + 100$$

dollars for a daily output of x tons of chocolate. The total revenue function is

$$R(x) = 1200x - 100x^2$$

dollars for an output of x tons a day. The marginal revenue and marginal cost functions are

$$R'(x) = 1200 - 200x$$
$$c'(x) = 2x + 2.$$

Setting $R'(x)$ equal to $c'(x)$ and solving for x yields

$$1200 - 200x = 2x + 2$$
$$202x = 1198$$
$$x = 5.93.$$

Thus $x = 5.93$ is a critical value. Now $R''(x) = -200$, so that $R''(5.93) = -200$. Also, $c''(x) = 2$, so that $c''(5.93) = 2$. Since $R''(5.93) < c''(5.93)$, it follows that $x = 5.93$ is the output which yields the maximum profit. The maximum profit is the difference between total revenue and total cost for $x = 5.93$,

$$P(5.93) = R(5.93) - c(5.93) = \$3453$$

which agrees with the results obtained in Example 2, Section 20.

Profit maximization after taxation

Let us suppose that the monopolist must pay the government a lump-sum profit tax T and examine how his profit-maximizing output is affected. His profit after taxes is

$$P(x) = R(x) - c(x) - T$$

where $R(x)$ and $c(x)$ are his total revenue and total cost functions. The derivative of $P(x)$ is

$$\frac{dP}{dx} = \frac{dR}{dx} - \frac{dc}{dx} - \frac{dT}{dx}.$$

$\frac{dT}{dx} = 0$ since T is a constant. Thus we have

$$\frac{dP}{dx} = \frac{dR}{dx} - \frac{dc}{dx}.$$

Setting $\frac{dP}{dx}$ equal to zero yields

$$\frac{dR}{dx} - \frac{dc}{dx} = 0$$

from which we obtain

$$\frac{dR}{dx} = \frac{dc}{dx}.$$

Therefore the profit-maximizing output of the monopolist is not altered by the imposition of a lump-sum tax T. It is still that output at which marginal revenue equals marginal cost. However, the monopolist's total profit is reduced by amount T.

For our chocolate-producing monopolist (see Example 5), this means that the optimal chocolate output remains at 5.93 tons a day, but the daily profit of \$3453 is reduced by amount T.

Suppose the monopolist must pay the government a fixed proportion t $(0 < t < 1)$ of the difference between his total revenue and total cost. How does this affect his profit-maximizing output?

The monopolist's profit after taxes is

$$P(x) = R(x) - c(x) - t[R(x) - c(x)]$$

where $R(x)$ and $c(x)$ are his total revenue and total cost functions. Factoring out $[R(x) - c(x)]$ yields

$$P(x) = (1 - t)[R(x) - c(x)].$$

The derivative of $P(x)$ is

$$\frac{dP}{dx} = (1 - t)\left[\frac{dR}{dx} - \frac{dc}{dx}\right].$$

Setting $\frac{dP}{dx}$ equal to zero yields

$$(1 - t)\left[\frac{dR}{dx} - \frac{dc}{dx}\right] = 0.$$

Since $(1 - t) \neq 0$, it follows that

$$\frac{dR}{dx} - \frac{dc}{dx} = 0$$

and thus

$$\frac{dR}{dx} = \frac{dc}{dx}.$$

In this situation, too, the profit-maximizing output of the monopolist is not altered by the imposition of a tax which is a fixed proportion of the difference between his total revenue and total cost.

Finally, let us suppose that an output tax of t monetary units per unit output is imposed on the monopolist. Then his profit function becomes

$$P(x) = R(x) - c(x) - tx$$

where $R(x)$ and $c(x)$ are his total revenue and total cost functions. The derivative of $P(x)$ is

$$\frac{dP}{dx} = \frac{dR}{dx} - \frac{dc}{dx} - t.$$

Setting $\frac{dP}{dx}$ equal to zero yields

$$\frac{dR}{dx} - \frac{dc}{dx} - t = 0$$

from which we obtain

$$\frac{dR}{dx} = \frac{dc}{dx} + t.$$

Therefore, if an output tax of t monetary units per unit output is to be paid, the monopolist's profit-maximizing output is that output at which marginal revenue equals marginal cost plus the unit tax.

In many situations a lump-sum tax is preferable to an output tax. To see why this is the case in specific terms, let us again return to our chocolate-producing monopolist.

EXAMPLE 6. The total revenue and total cost functions of a chocolate-producing monopolist are

$$R(x) = 1200x - 100x^2$$
$$c(x) = x^2 + 2x + 100.$$

$R(x)$ expresses the total dollar revenue derived from an output of x tons of chocolate a day and $c(x)$ expresses the total dollar cost of producing x tons of chocolate a day. The profit function $P(x)$ is

$$P(x) = R(x) - c(x) = -101x^2 + 1198x - 100.$$

The demand function

$$p = 60 - 5x$$

expresses p, the market price of chocolate in cents per pound, as a function of x, the output in tons per day.

In Example 5 we saw that, before taxes, the profit-maximizing output is 5.93 tons of chocolate a day, with a corresponding profit of $P(5.93) = \$3453$. The market price corresponding to an output of 5.93 tons a day is $p = 60 - 5(5.93) = 30.35$ cents per pound. Let us suppose that an output tax of \$188 per ton is imposed. Then profit is maximized for that output at which

$$\frac{dR}{dx} = \frac{dc}{dx} + 188.$$

Since $\frac{dR}{dx} = 1200 - 200x$ and $\frac{dc}{dx} = 2x + 2$, we obtain

$$1200 - 200x = 2x + 2 + 188$$
$$202x = 1010$$
$$x = 5.$$

Thus profit-maximizing output is reduced from 5.93 tons per day to 5 tons per day and the corresponding market price of chocolate rises from 30.35 cents per pound to 35 cents per pound. The tax revenue derived by the government is $5(\$188) = \$940$ and the monopolist's profit after taxes is $P(5) - 940 = \$3365 - \$940 = \$2425$.

A lump-sum tax of \$940 per day is equivalent to an output tax of \$188 per ton a day and is preferable in many respects. The government obtains the same tax revenue, but the daily optimal output remains at 5.93 tons and does not drop to 5 tons, market price stays at 30.35 cents per pound, and the monopolist's profit after taxes is higher: $P(5.93) - 940 = \$2513$ as opposed to $P(5) - 940 = \$2425$.

Exercises

33. A sugar refinery produces x tons of sugar a week at a total cost of $c(x) = \frac{1}{10}x^2 + 2x + 200$ dollars and with a total revenue of $R(x) = 102x - \frac{9}{10}x^2$ dollars.
    a. What is the sugar-producing monopolist's optimal output and profit before tax considerations?
    b. What effect does an output tax of $40 per ton have on optimal output?
    c. What revenue can the tax collector expect to obtain?
    d. What is the monopolist's profit after taxes?
    e. If a lump-sum tax equivalent to an output tax of $40 a ton is imposed, what is the monopolist's profit after taxes?
    f. The demand function for sugar is $p = 102 - \frac{9}{10}x$, where p is the market price in dollars per ton and x is output in tons per week. What effect does an output tax of $40 per ton have on the market price?

34. The output of a coffee-producing monopolist is x tons a week at a total cost of $c(x) = x^2 + 4x + 3000$ dollars and with a total revenue of $R(x) = 1204x - 5x^2$ dollars.
    a. Determine the coffee-producing monopolist's optimal output and profit before tax considerations.
    b. What effect does an output tax of $120 per ton have on optimal output?
    c. What revenue can the tax collector expect to obtain?
    d. What is the monopolist's profit after taxes?
    e. If a lump-sum tax equivalent to an output tax of $120 a ton is imposed, what is the monopolist's profit after taxes?
    f. The demand function for coffee is $p = 1204 - 5x$, where p is the market price in dollars per ton and x is output in tons per week. What effect does an output tax of $120 per ton have on market price?

35. A tobacco monopolist produces x tons a week at a total cost of $c(x) = \frac{1}{4}x^2 + \frac{3}{2}x + 2000$ dollars. The demand function for tobacco is $p = 600 - 5x$, where p is the market price in dollars per ton and x is output in tons per week.
    a. Determine the tobacco monopolist's optimal output and profit before tax considerations.
    b. What effect does an output tax of $73.50 per ton have on optimal output?
    c. What revenue can the tax collector expect to obtain?
    d. What is the monopolist's profit after taxes?
    e. If a lump-sum tax equivalent to an output tax of $73.50 a ton is imposed, what is the monopolist's profit after taxes?
    f. What effect does an output tax of $73.50 per ton have on market price?

36. The output of an aluminum-producing monopolist is x tons a week at a total cost of $c(x) = \frac{1}{3}x^3 - 9x^2 + 100x + 1000$ dollars. The demand function for

aluminum is $p = 1060 - x$, where p is the market price in dollars per ton and x is output in tons per week.

a. Determine the aluminum monopolist's optimal output and profit before tax considerations.
b. What effect does an output tax of $295 per ton have on optimal output?
c. What revenue can the tax collector expect to obtain?
d. What is the monopolist's profit after taxes?
e. If a lump-sum tax equivalent to an output tax of $295 a ton is imposed, what is the monopolist's profit after taxes?
f. What effect does an output tax of $295 per ton have on market price?

22. Tax revenue maximization

Let us return once more to our chocolate-producing monopolist, but this time let us examine his situation from the point of view of the tax collector who wishes to maximize tax revenue. In Example 6, Section 21, we saw that if an output tax of $188 per ton is imposed, then the total, daily, tax revenue derived is $940, based on a daily, profit-maximizing output of 5 tons. Is this the best the government can do?

The question that is naturally raised is this: What tax rate t should be used so as to maximize tax revenue? To answer this question we must determine the daily output for which profit is maximized after taxes. In the previous section we saw that the monopolist's profit-maximizing output after taxes is that output at which marginal revenue is equal to marginal cost plus the unit tax, that is,

$$\frac{dR}{dx} = \frac{dc}{dx} + t.$$

Since the total revenue and total cost functions of the monopolist are

$$R(x) = 1200x - 100x^2$$
$$c(x) = x^2 + 2x + 100$$

his marginal revenue and marginal cost functions are

$$R'(x) = 1200 - 200x$$
$$c'(x) = 2x + 2.$$

Let marginal revenue equal marginal cost plus the tax rate t and solve for x.

$$1200 - 200x = 2x + 2 + t$$
$$202x = 1198 - t$$
$$x = \frac{1198}{202} - \frac{t}{202}$$
$$x = 5.93 - \frac{t}{202}$$

Therefore the output at which profit is maximized after taxes is

$$x = 5.93 - \frac{t}{202}.$$ (1)

Consider the tax-revenue function

$$T(t) = t\left(5.93 - \frac{t}{202}\right) = 5.93t - \frac{t^2}{202}$$

obtained by multiplying tax rate t by the output at which the monopolist's profit is maximized after the imposition of tax rate t. Our problem is to determine the value of t for which this tax-revenue function is maximized. To do so, we find $T'(t)$, set it equal to zero, and solve for t.

$$T'(t) = 5.93 - \frac{2t}{202} = 5.93 - \frac{t}{101}$$

$$5.93 - \frac{t}{101} = 0$$

$$\frac{t}{101} = 5.93$$

$$t = 598.93$$

Now $T''(t) = -\frac{1}{101}$, so $T''(t) < 0$ for all t. Therefore $T''(598.93) < 0$, and it follows by the second derivative test that $t = 598.93$ maximizes our tax-revenue function.

Thus an output tax of $598.93 should be imposed on each ton of chocolate produced to maximize tax revenue. By substituting 598.93 for t in equation (1) and solving for x, we obtain $x = 2.96$ tons a day as the chocolate monopolist's profit-maximizing output after taxes. The daily tax revenue derived by the government is $(2.96)(598.93) = \$1773$.

Since the chocolate producer's profit function before taxes is

$$P(x) = -101x^2 + 1198x - 100$$

then his daily profit after taxes is

$$P(2.96) - 1773 = \$2566 - \$1773 = \$793.$$

Since the demand function for chocolate is

$$p = 60 - 5x$$

then the market price of chocolate after the imposition of the output tax would rise to

$$p = 60 - 5(2.96) = 45.20 \text{ cents per pound.}$$

The lump-sum tax of $1773 a day is equivalent to an output tax of $598.93 per ton produced per day and avoids some of its negative features. The tax revenue obtained by the government is the same but daily optimal output is 5.93 tons instead of 2.96 tons, market price is 30.35 cents per pound instead of 45.20 cents per pound, and the chocolate producer's daily profit after taxes is $P(5.93) - \$1773 = \$3453 - \$1773 = \$1680$ instead of $793.

Exercises

37. Refer to the sugar-refinery situation discussed in Exercise 33.
    a. What is the output tax rate that should be used to maximize tax revenue?
    b. What effect does this output tax have on optimal output?
    c. What revenue can the tax collector expect to obtain?
    d. What is the monopolist's profit after taxes?
    e. If a lump-sum tax equivalent to this output tax is imposed, what is the monopolist's profit after taxes?
    f. What effect does this output tax have on the market price of sugar?

38. Refer to the coffee-production situation discussed in Exercise 34.
    a. What is the output tax rate that should be used to maximize tax revenue?
    b. What effect does this output tax have on optimal output?
    c. What revenue can the tax collector expect to obtain?
    d. What is the monopolist's profit after taxes?
    e. If a lump-sum tax equivalent to this output tax is imposed, what is the monopolist's profit after taxes?
    f. What effect does this output tax have on the market price of coffee?

39. Refer to the tobacco-production situation discussed in Exercise 35.
    a. What is the output tax rate that should be used to maximize tax revenue?
    b. What effect does this output tax have on optimal output?
    c. What revenue can the tax collector expect to obtain?
    d. What is the monopolist's profit after taxes?
    e. If a lump-sum tax equivalent to this output tax is imposed, what is the monopolist's profit after taxes?
    f. What effect does this output tax have on the market price of tobacco?

23. An optimal storage time problem

At a certain time, which becomes our time point of reference, a certain commodity is obtained. It is stored and it increases in value as time passes. The selling price of the commodity x years after its initial introduction is a known function,

$$S = f(x)$$

where x is the time elapsed and S is the selling price per unit at time x. Wine and timber are classic examples of commodities whose values increase with time. New wine is bought by a dealer to be sold at a later time. An investor acquires land on which timber has been planted and which is to be sold at a later time. The problem is to determine the optimal time at which to sell the commodity.

Specifically, suppose timber-bearing land has been obtained, and

$$S = 1000 \, e^{\sqrt{x}}$$

expresses S, the dollar selling price of the land, as a function of x, the number of years after its acquisition. Before we can formulate our optimization problem, a question of interest must be considered. The problem is that the sale value at one time is not directly comparable to the sale value at another time. Because of the time lag, growth due to interest accumulation must be taken into account.

Let us assume that interest accumulates at a rate r compounded continuously. To be very specific, suppose the problem is to compare a sale value of $2000 two years after acquisition of a commodity with a sale value of $3000 three years after the acquisition of the commodity, where interest is accumulated at a rate of 5% compounded continuously. Then the two sums can be compared by reference to the amounts that must initially be invested at a rate of 5% compounded continuously if at the end of two years $2000 is to be attained and if at the end of three years $3000 is to be attained.

In Section 7 we saw that

$$A\, e^{-rx}$$

expresses the amount that must initially be invested at rate r compounded continuously if at the end of x years sum A is to be attained. Thus

$$2000\, e^{-0.05(2)} = 2000(0.9048) = \$1809.60$$

grows to a value of $2000 in two years, and

$$3000\, e^{-0.05(3)} = 3000(0.8607) = \$2582.10$$

grows to a value of $3000 in three years. Therefore, if the stated interest conditions are taken into consideration, a comparison of a sale value of $2000 after two years with a sale value of $3000 after three years is equivalent to comparing $1809.60 with $2582.10.

Thus returning to our timber-land situation, if

$$S = 1000\, e^{\sqrt{x}}$$

expresses the dollar sale value of the land as a function of the number of years after its acquisition and interest is accumulated at a rate of r compounded continuously, then

$$y = S\, e^{-rx} = 1000\, e^{\sqrt{x}}\, e^{-rx} = 1000\, e^{\sqrt{x}-rx}$$

expresses the value of the timber land x years after its acquisition when the given interest conditions have been taken into consideration. Our optimization problem is to determine the value of x for which y is a maximum. To do this we must determine $\dfrac{dy}{dx}$, set it equal to zero, and solve for x. By using the chain rule we obtain

$$\frac{dy}{dx} = 1000\, e^{\sqrt{x}-rx}\left(\frac{1}{2\sqrt{x}} - r\right).$$

Setting $\dfrac{dy}{dx}$ equal to zero yields

$$1000\, e^{\sqrt{x}-rx}\left(\frac{1}{2\sqrt{x}} - r\right) = 0.$$

Since $1000\, e^{\sqrt{x}-rx} \neq 0$, this condition can only be satisfied if

$$\frac{1}{2\sqrt{x}} - r = 0.$$

Solving for x yields

$$x = \frac{1}{4r^2}.$$

To see if $x = 1/4r^2$ yields a maximum value, we use the first derivative test. If $x < 1/4r^2$,

$$\frac{dy}{dx} = \underbrace{1000\, e^{\sqrt{x}-rx}}_{\text{positive}} \underbrace{\left(\frac{1}{2\sqrt{x}} - r\right)}_{\text{positive}}.$$

$\dfrac{dy}{dx}$ is the product of two positive factors and is thus positive. If $x > 1/4r^2$,

$$\frac{dy}{dx} = \underbrace{1000\, e^{\sqrt{x}-rx}}_{\text{positive}} \underbrace{\left(\frac{1}{2\sqrt{x}} - r\right)}_{\text{negative}}.$$

$\dfrac{dy}{dx}$ is the product of a positive factor and a negative factor and is thus negative. Since $\dfrac{dy}{dx}$ goes from positive to negative, it follows from the first derivative test that

$$y = 1000\, e^{\sqrt{x}-rx}$$

has a maximum value at $x = 1/4r^2$. Therefore the higher the interest rate is, the shorter the optimal waiting time will be.

Thus, for example, if interest is accumulated at the rate of 10% compounded continuously, the optimal time to sell the timber is

$$x = \frac{1}{4(0.10)(0.10)} = 25 \text{ years}$$

after its acquisition. Its value after 25 years is

$$y = 1000\, e^{(\sqrt{25}-0.1(25))} = 1000\, e^{2.5} = 1000(12.1825) = \$12{,}182.50.$$

We should also note that while our formulation of the sale-time optimization problem takes interest considerations into account, it does not take storage or maintenance costs into account.

Exercises

40. A wine merchant has purchased a quantity of wine which is to be aged and resold at a later time. The function

$$S = 500\, e^{\frac{1}{2}\sqrt{x}}$$

expresses S, the dollar selling price of the wine, as a function of x, the number of years after its acquisition. Interest accumulates at the rate of 10% compounded continuously. Determine the optimal storage time of the wine.

41. An investor has purchased timber-bearing land which is to be resold at a later time. The function

$$S = 2000\, e^{\sqrt{x}-0.1x}$$

expresses the selling price of the land as a function of the number of years after its acquisition. For interest accumulating at the rate of 10% compounded continuously, determine the optimal time to sell the timber.

24. Economic relationships at extreme values

Marginal and average costs

Consider a firm producing a single, uniform commodity whose total cost for output x (per unit time) is described by the differentiable function

$$c = f(x), \quad x > 0.$$

A proposition of economics, which we will prove here, is that the marginal cost curve of the firm, $M_c = f'(x)$, intersects the average cost curve, $\bar{c} = f(x)/x$, at its minimum point.

Proof. Part of the background provided by economic theory is that the average cost function $\bar{c} = f(x)/x$ has exactly one minimum point. If we let a denote the output at which the minimum value occurs, then the coordinates of the minimum point are $(a, f(a)/a)$. Since a gives rise to an extreme value, it is a critical value of the average cost function and thus must satisfy one of the following three conditions.

1. $\bar{c}'(a) = 0$.
2. $\bar{c}'(a)$ does not exist.
3. a is an end point of the domain of definition of the average cost function.

Conditions 2 and 3 are ruled out by circumstances. It is assumed that the domain of definition of $\bar{c}(x)$ does not have end points ($x > 0$) and that $\bar{c}'(x)$ is defined wherever $\bar{c}(x)$ is defined. Thus a must satisfy the condition

$$\bar{c}'(a) = 0.$$

Since $\bar{c}(x) = f(x)/x$ is a quotient whose components have derivatives, we can use the quotient theorem to determine $\bar{c}'(a)$. We obtain

$$\bar{c}'(a) = \frac{a \cdot f'(a) - f(a)}{a^2}.$$

Since $\bar{c}'(a) = 0$, we have

$$a \cdot f'(a) - f(a) = 0$$

so that

$$\underbrace{f'(a)}_{\substack{\text{marginal} \\ \text{cost for} \\ \text{output } a}} = \underbrace{\frac{f(a)}{a}}_{\substack{\text{average} \\ \text{cost for} \\ \text{output } a}}$$

Thus a is the output at which marginal cost equals average cost. Therefore $(a, f'(a)) = (a, f(a)/a)$, which means that the marginal cost curve intersects the average cost curve at its minimum point.

Elasticity of Demand and Total Revenue

Consider a firm producing a single, uniform commodity for which the demand function is

$$x = f(p), \quad p > 0$$

where x is the quantity purchased by the market per unit time and p is the unit price of the commodity. A proposition of economics, which we will prove here, is that elasticity of demand η is 1 at that value of p which maximizes total revenue.

Proof. The firm's total revenue obtained from the sale of $x = f(p)$ units at price p per unit is

$$R = px = p \cdot f(p).$$

Part of the background provided by economic theory is that the revenue function $R = p \cdot f(p)$ has exactly one maximum value. Let a denote the value of p at which this maximum value occurs. Now let us recall that the elasticity of demand for price a is defined by

$$\eta = -\frac{a}{f(a)} f'(a).$$

Thus we must show that

$$-\frac{a}{f(a)} f'(a) = 1.$$

Since a is a critical value for $R = p \cdot f(p)$, it must satisfy one of the following three conditions.

1. $R'(a) = 0$.
2. $R'(a)$ does not exist.
3. a is an end point of the domain of definition of $R = p \cdot f(p)$.

Conditions 2 and 3 are ruled out by the circumstances of the situation, and thus $R'(a)$ must satisfy the condition

$$R'(a) = 0.$$

Since $R = p \cdot f(p)$ is a product whose components have derivatives, we can use the product theorem to obtain an expression for $R'(a)$.

$$R'(a) = a \cdot f'(a) + f(a)$$

Since $R'(a) = 0$, we obtain the following results.

$$a \cdot f'(a) + f(a) = 0$$
$$a \cdot f'(a) = -f(a)$$

Dividing both members by $-f(a)$ yields

$$-\frac{a}{f(a)} f'(a) = 1.$$

Our proposition is established.

4

Curve sketching

25. Curve sketching: the calculus to the rescue

Let us recall that the graph of a function $y = f(x)$ is the collection of points with coordinates $(x, f(x))$, where x takes on only those values in its domain of definition. The visual image of a function, its graph, provides us with an extremely useful tool in helping us understand the function's behavior.

One approach to graphing functions, which must be used with discretion, is to let the independent variable take on a number of values, calculate the corresponding values of the dependent variable, plot the corresponding points, and join these points with a smooth curve. However, as we have seen in the previous chapters, this approach can lead to errors. In this chapter we will examine a number of powerful aids to graph sketching which are provided by the calculus.

In order to avoid making errors when we are joining the plotted points with a smooth curve, we must be sure to look for jumps, gaps, and breaks in the graph. Thus the question of continuity arises since the graph of a function continuous over an interval has no jumps, gaps, or breaks in that interval. Recall that a function is continuous wherever it has a derivative.* To illustrate, consider the average cost function

$$\bar{c}(x) = x + 2 + \frac{100}{x}, \qquad x > 0$$

---

*At the same time, remember that a function which does not have a derivative at a value may still be continuous at the value (see Section 10). To settle the continuity issue for a function which does not have a derivative at a value, we must use the definition of continuity.

116

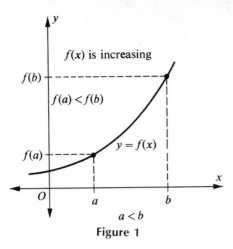

Figure 1

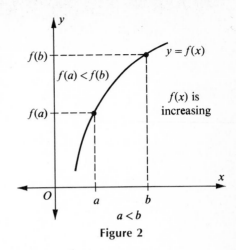

Figure 2

for a chocolate producer whose output is x tons of chocolate per day (see Example 1, Section 20). $\bar{c}'(x)$ exists for all $x > 0$ and is given by

$$\bar{c}'(x) = 1 - \frac{100}{x^2}, \qquad x > 0.$$

Thus $\bar{c}(x)$ is continuous for all $x > 0$, and its graph has no jumps, gaps, or breaks.

Knowledge of a function's extreme values is very useful in sketching its graph. Knowing the local maxima or local minima gives us some idea of what the curve looks like at and near those values. In Example 1, Section 20, we established that the chocolate producer's average cost function $\bar{c}(x)$ has a local minimum value at $x = 10$. Thus $(10, \bar{c}(10))$, or $(10, 22)$, is a minimum point.

Another aid in graph sketching is determination of where the function is increasing and where it is decreasing. Figures 1 and 2 illustrate functions which are increasing over an interval, and Figures 3 and 4 illustrate functions which are decreasing over

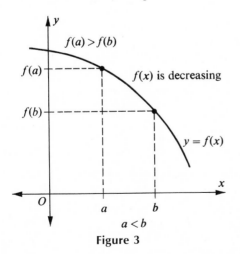

Figure 3

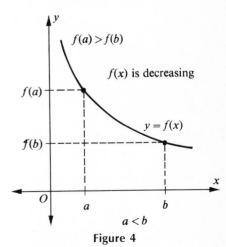

Figure 4

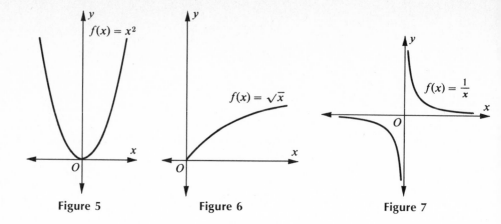

Figure 5   Figure 6   Figure 7

an interval. As these graphs make clear, a function $y = f(x)$ is *increasing* in an interval if for any two values a and b in the interval, $f(a) < f(b)$ whenever $a < b$. A function $y = f(x)$ is *decreasing* in an interval if for any two values a and b in the interval, $f(a) > f(b)$ whenever $a < b$. In addition to intervals, these definitions hold for half-lines and the entire x-axis.

$f(x) = x^2$, shown in Figure 5, is increasing for $x \geq 0$ and is decreasing for $x \leq 0$. $f(x) = \sqrt{x}$, shown in Figure 6, is increasing for $x \geq 0$. $f(x) = 1/x$, shown in Figure 7, is decreasing for $x < 0$ and is also decreasing for $x > 0$.

The calculus provides us with a simple method for determining where a function $y = f(x)$ is increasing and decreasing.

Test for Increasing and Decreasing Behavior. If $f'(x) > 0$ in an interval, then $y = f(x)$ is *increasing* in the interval. If $f'(x) < 0$ in an interval, then $y = f(x)$ is *decreasing* in the interval.

In addition to intervals, this result also holds for such regions as half-lines and the entire x-axis.

As an illustration within a familiar setting, consider $f(x) = x^2$, which is decreasing for $x \leq 0$ and increasing for $x \geq 0$ (see Figure 5). The derivative of $f(x) = x^2$ is

$$f'(x) = 2x$$

which is negative for $x < 0$ and positive for $x > 0$.

As we have observed, $f(x) = 1/x$ (shown in Figure 7) is decreasing for $x < 0$ and is also decreasing for $x > 0$. The derivative of $f(x) = 1/x$ is

$$f'(x) = -\frac{1}{x^2}$$

which is negative for $x < 0$ and negative for $x > 0$.

For the average cost function

$$\bar{c}(x) = x + 2 + \frac{100}{x}, \qquad x > 0$$

the derivative is

$$\bar{c}'(x) = 1 - \frac{100}{x^2} = \frac{x^2 - 100}{x^2} = \frac{(x - 10)(x + 10)}{x^2}$$

where $x > 0$. Thus we see that $\bar{c}'(10) = 0$. For $x > 10$, we have

$$\bar{c}'(x) = \frac{\overbrace{(x - 10)}^{\text{pos.}}\overbrace{(x + 10)}^{\text{pos.}}}{x^2}$$

which is positive. For $x < 10$, where x is positive, we have

$$\bar{c}'(x) = \frac{\overbrace{(x - 10)}^{\text{neg.}}\overbrace{(x + 10)}^{\text{pos.}}}{x^2}$$

which is negative. In summary, then, $\bar{c}'(10) = 0$, $\bar{c}'(x) > 0$ for $x > 10$, and $\bar{c}'(x) < 0$ for $x < 10$, where x is positive. Thus $\bar{c}(x)$ is increasing for $x > 10$ and decreasing for $0 < x < 10$. Since we have already observed that $c(10)$ is a minimum value, this does not come as a surprise but rather as confirmation of information already obtained.

From Figures 1 and 2 we see that a distinction can be made in the manner in which a function is increasing. The distinction lies in what is called the concavity of the function. The graph of the function shown in Figure 8 is said to be *concave*

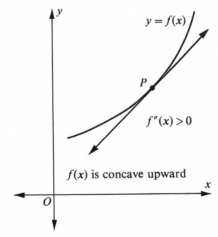

Figure 8

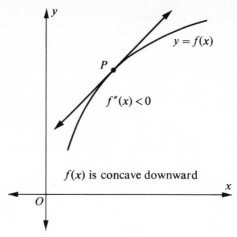

Figure 9

upward at point P since the graph is above its tangent line at P. The graph of the function shown in Figure 9 is below its tangent line at P and is thus said to be *concave downward* at P. The graph of the function shown in Figure 3 is concave downward at each point, while the graph of the function shown in Figure 4 is concave upward at each point. The graph of $f(x) = x^2$, shown in Figure 5, is concave upward at each point, while the graph of $f(x) = \sqrt{x}$, shown in Figure 6, is concave downward at each point. The graph of $f(x) = 1/x$, shown in Figure 7, is concave downward for $x < 0$ and concave upward for $x > 0$.

The concavity of a function $y = f(x)$ is related to the behavior of its second derivative in the following way.

Concavity Test. If $f''(x) > 0$ at each point of an interval, then the graph of $y = f(x)$ is *concave upward* at each point of the interval. If $f''(x) < 0$ at each point of an interval, then the graph of $y = f(x)$ is *concave downward* at each point of the interval.

In addition to intervals, this test holds for such regions as half-lines and the entire x-axis.

If a point $P(c, f(c))$ is such that $f''(c) = 0$ and $f''(x)$ goes from positive to negative, or negative to positive, as we go from $x < c$ to $x > c$, then $P(c, f(c))$ is a point which separates portions of the graph which are concave upward and concave downward. Such points, called *inflection points* of the graph, are often central to determining the concavity behavior of a function and are thus of considerable value in graph sketching.

To illustrate the concavity test within a familiar setting, consider again $f(x) = x^2$, whose graph is concave upward at all points (see Figure 5). $f'(x) = 2x$ and

$$f''(x) = 2$$

which is positive for all x. Hence there is no inflection point.

$f(x) = \sqrt{x}$, whose graph is concave downward for $x > 0$ (see Figure 6), has first derivative

$$f'(x) = \frac{1}{2}x^{-\frac{1}{2}}.$$

The second derivative is

$$f''(x) = -\frac{1}{4}x^{-\frac{3}{2}} = -\frac{1}{4\sqrt{x^3}}$$

which is negative for $x > 0$. Again, there is no inflection point.

$f(x) = 1/x$, whose graph is concave downward for $x < 0$ and concave upward for $x > 0$ (see Figure 7), has first derivative

$$f'(x) = -\frac{1}{x^2}.$$

The second derivative is

$$f''(x) = \frac{2}{x^3}$$

which is negative for $x < 0$ and positive for $x > 0$.

The average cost function $\bar{c}(x) = x + 2 + (100/x)$, where $x > 0$, has first derivative

$$\bar{c}'(x) = 1 - \frac{100}{x^2}, \qquad x > 0$$

and second derivative

$$\bar{c}''(x) = \frac{200}{x^3}, \qquad x > 0.$$

Since $\bar{c}''(x)$ is positive for $x > 0$, the graph of $\bar{c}(x)$ is concave upward for $x > 0$.

The function

$$f(x) = \frac{1}{6}x^3 - \frac{1}{2}x^2 - 4x + 2$$

has first derivative

$$f'(x) = \frac{1}{2}x^2 - x - 4$$

and second derivative

$$f''(x) = x - 1.$$

The first step in determining points of inflection is to set $f''(x)$ equal to zero and solve for x. We obtain

$$x - 1 = 0$$
$$x = 1.$$

Thus $f''(1) = 0$. For $x < 1$, $f''(x) < 0$, and for $x > 1$, $f''(x) > 0$. Thus $(1, f(1))$, or $\left(1, -\dfrac{7}{6}\right)$, is a point of inflection. The graph of $f(x)$ is concave downward for $x < 1$ and concave upward for $x > 1$.

Exercises

Determine where the following functions are increasing and decreasing, the points of inflection, and concavity behavior.

1. $f(x) = x^3 + 1$

2. $\bar{c}(x) = \dfrac{1}{4}x + \dfrac{3}{2} + \dfrac{1}{x}, \quad x > 0$

3. $\bar{c}(x) = \dfrac{1}{100}x + 3 + \dfrac{100}{x}, \quad x > 0$

4. $P(x) = -x^2 + 900x - 10, \quad x > 0$

5. $P(x) = -\dfrac{1}{3}x^3 + 7x^2 + 1800x - 100, \quad x > 0$

6. $f(x) = x^3 + \frac{7}{2}x^2 - 6x - 10$

7. $f(x) = \frac{2}{3}x^3 - \frac{9}{2}x^2 + 4x - 12$

8. $f(x) = \frac{1}{3}x^3 - \frac{1}{2}x^2 - 6x - 4$

9. $f(x) = xe^x$

10. $f(x) = xe^{-x}$

In sketching the graph of a function, it is sometimes useful to explore the limit behavior of the function as x approaches certain values and as $x \to \infty$ and as $x \to -\infty$. For example, since $\bar{c}(x) = x + 2 + (100/x)$ has a term with x in the denominator, the behavior of $\bar{c}(x)$ as x approaches 0 might prove of interest. Let us observe that as $x \to 0$, where $x > 0$, $\bar{c}(x) \to \infty$. Let us also observe that as $x \to \infty$, $\bar{c}(x) \to \infty$.

Knowledge of the x-intercepts and y-intercept of the curve, that is, the points at which the graph intersects the x-axis ($y = 0$) and the y-axis ($x = 0$), are also sometimes useful. For example, since $\bar{c}(x) = x + 2 + (100/x)$, $\bar{c}(x) > 0$ for $x > 0$, and $\bar{c}(x)$ is entirely above the x-axis.

In summary, then, knowledge of the following is of considerable value in sketching the graph of a function.

1. Continuity.
2. Extreme values.
3. Regions in which the function is increasing and decreasing.
4. Points of inflection and the concavity behavior of the function.
5. Limit behavior with respect to certain values and as $x \to \infty$ and as $x \to -\infty$.
6. x-intercepts and y-intercept.

EXAMPLE 1. Sketch the graph of the chocolate producer's average cost function

$$\bar{c}(x) = x + 2 + \frac{100}{x}, \; x > 0.$$

Solution. We have the following results for $\bar{c}(x)$ as discussed in this section.

1. $\bar{c}(x)$ is continuous for all $x > 0$.
2. $(10, 22)$ is a minimum point.
3. $\bar{c}(x)$ decreases for $0 < x < 10$ and increases for $x > 10$.
4. $\bar{c}(x)$ is concave upward for all $x > 0$. There are no inflection points.
5. $\lim\limits_{x \to 0} \bar{c}(x) = \infty$, $\lim\limits_{x \to \infty} \bar{c}(x) = \infty$.
6. $\bar{c}(x)$ is entirely above the x-axis.

In the following table, values for $\bar{c}(x)$ are given for certain values of x.

x	1	5	10	20	50	100
$\bar{c}(x)$	103	27	22	27	54	103

We thus obtain the graph shown in Figure 10.

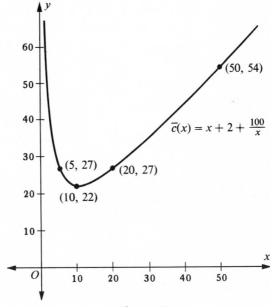

Figure 10

EXAMPLE 2. Sketch the graph of $f(x) = 2x^3 + 3x^2 - 12x + 1$.

Solution. 1. Continuity. Since $f'(x)$ exists for all x, $f(x)$ is continuous at all real numbers. Thus the graph of $f(x)$ has no jumps, gaps, or breaks.

2. Extreme values. There are no end point critical values and no values at which $f'(x)$ does not exist. To obtain type 1 critical values, we must determine $f'(x)$, set it equal to zero, and solve for x.

$$f'(x) = 6x^2 + 6x - 12$$

Setting $f'(x)$ equal to zero yields

$$6x^2 + 6x - 12 = 0$$
$$6(x^2 + x - 2) = 0$$
$$6(x + 2)(x - 1) = 0$$
$$x = -2, \quad x = 1.$$

Thus $x = -2$ and $x = 1$ are type 1 critical values. To determine if they yield extreme values, we use the second derivative test.

$$f''(x) = 12x + 6$$
$$f''(-2) = -18, \quad f''(1) = 18$$

Since $f''(-2) < 0$, $f(-2) = 21$ is a local maximum value. $f''(1) > 0$, and thus $f(1) = -6$ is a local minimum value. Therefore $(-2, 21)$ is a local maximum point and $(1, -6)$ is a local minimum point.

3. Regions in which $f(x)$ increases and decreases. To find where $f(x)$ increases and decreases, we must determine where $f'(x)$ is positive and negative.

$$f'(x) = 6(x + 2)(x - 1)$$

Since $f'(x) = 0$ for $x = -2$ and $x = 1$, we examine the behavior of $f'(x)$ for $x < -2$, $-2 < x < 1$, and $x > 1$.

For $x < -2$,

$$f'(x) = \underbrace{6}_{\text{pos.}} \ \underbrace{(x + 2)}_{\text{neg.}} \underbrace{(x - 1)}_{\text{neg.}}.$$

$f'(x)$ is the product of a positive factor and two negative factors and is thus positive. Therefore $f(x)$ is increasing for $x < -2$.

For $-2 < x < 1$,

$$f'(x) = \underbrace{6(x + 2)}_{\text{pos.}}\underbrace{(x - 1)}_{\text{neg.}}.$$

$f'(x)$ is the product of a positive factor and a negative factor and is thus negative. Therefore $f(x)$ is decreasing for $-2 < x < 1$.

For $x > 1$,

$$f'(x) = \underbrace{6(x + 2)}_{\text{pos.}}\underbrace{(x - 1)}_{\text{pos.}}.$$

$f'(x)$ is a product of positive factors and is thus positive. Therefore $f(x)$ is increasing for $x > 1$.

4. Points of inflection and concavity. The first step in finding points of inflection is determining where $f''(x) = 0$.

$$f''(x) = 12x + 6 = 12\left(x + \frac{1}{2}\right)$$

Setting $f''(x)$ equal to zero and solving for x yields

$$12\left(x + \frac{1}{2}\right) = 0$$

$$x = -\frac{1}{2}.$$

Thus $f''(-\frac{1}{2}) = 0$. For $x < -\frac{1}{2}$, $f''(x) < 0$, and for $x > -\frac{1}{2}$, $f''(x) > 0$. Thus $(-\frac{1}{2}, f(-\frac{1}{2}))$, or $(-\frac{1}{2}, \frac{15}{2})$, is a point of the inflection. The graph of $f(x)$ is concave downward for $x < -\frac{1}{2}$ and concave upward for $x > -\frac{1}{2}$.

5. Limit behavior. As $x \to \infty$, $f(x) \to \infty$, and as $x \to -\infty$, $f(x) \to -\infty$.

In summary, then, the graph of $f(x) = 2x^3 + 3x^2 - 12x + 1$ has no jumps, gaps, or breaks. $(-2, 21)$ is a local maximum point and $(1, -6)$ is a local minimum point. The graph increases for $x < -2$, decreases for $-2 < x < 1$, and increases for $x > 1$. $(-\frac{1}{2}, \frac{15}{2})$ is an inflection point and the graph is concave downward for $x < -\frac{1}{2}$ and concave upward for $x > -\frac{1}{2}$. As $x \to \infty$, $f(x) \to \infty$, and as $x \to -\infty$, $f(x) \to -\infty$.

In the following table, values for $f(x)$ are given for certain values of x.

x	-5	-4	-3	-1	0	2	3	5
$f(x)$	-114	-31	10	14	1	5	46	266

We thus obtain the graph shown in Figure 11 on the next page.

EXAMPLE 3. Sketch the graph of C. L. Hull's learning theory function, $h(x) = 100(1 - e^{-ax})$, $x \geq 0$, where a is a positive constant.

Solution. Let us begin our analysis by observing that $h'(x)$ exists for $x \geq 0$ and is given by

$$h'(x) = 100ae^{-ax}.$$

Since $h'(x)$ exists for $x \geq 0$, $h(x)$ is continuous for $x \geq 0$. Therefore the graph of $h(x)$ has no jumps, gaps, or breaks.

Setting $h'(x)$ equal to zero yields

$$100ae^{-ax} = 0.$$

Since $100a$ is a positive constant and e^{-ax} is positive for all x, $100ae^{-ax}$ is never zero and, in fact, is positive for $x \geq 0$. Thus there are no type 1 critical values. $h'(x) > 0$ implies that $h(x)$ is increasing. Since 0 is an end point critical value and $h(x)$ is increasing, it follows that $h(0) = 0$ is a minimum value.

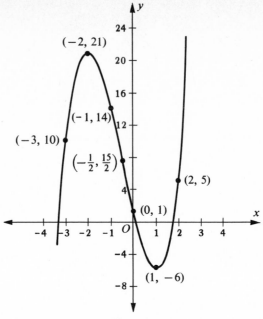

Figure 11

Now determine the derivative of $h'(x) = 100ae^{-ax}$.

$$h''(x) = -100a^2e^{-ax}$$

$h''(x)$ is negative for $x \geq 0$ since $100a^2$ is positive and e^{-ax} is positive. Thus there are no points of inflection (since $h''(x) \neq 0$), and the graph of $h(x)$ is concave downward for $x \geq 0$.

As $x \to \infty$, $\dfrac{1}{e^{ax}} \to 0$, so that $100(1 - e^{-ax}) \to 100$.

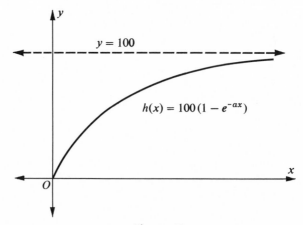

Figure 12

In summary, then, the graph of $h(x) = 100(1 - e^{-ax})$, $x \geq 0$, has no gaps, jumps. or breaks, is increasing, and is concave downward. As $x \to \infty$, $h(x) \to 100$. We therefore obtain the graph shown in Figure 12.

Exercises

Sketch the graphs of the following average cost functions.

11. $\bar{c}(x) = \dfrac{1}{10}x + 2 + \dfrac{10}{x}$, $x > 0$

12. $\bar{c}(x) = \dfrac{1}{4}x + \dfrac{3}{2} + \dfrac{1}{x}$, $x > 0$

13. $\bar{c}(x) = \dfrac{1}{100}x + 3 + \dfrac{100}{x}$, $x > 0$

Sketch the graphs of the following profit functions.

14. $P(x) = -x^2 + 900x - 10$, $x > 0$
15. $P(x) = -101x^2 + 1198x - 100$, $x > 0$
16. $P(x) = -x^2 + 1002x - 100$, $x > 0$
17. $P(x) = -\frac{1}{3}x^3 + 7x^2 + 1800x - 100$, $x > 0$

Sketch the graphs of the following functions.

18. $f(x) = x^3 + \frac{7}{2}x^2 - 6x - 10$
19. $f(x) = \frac{2}{3}x^3 - \frac{9}{2}x^2 + 4x - 12$
20. $f(x) = \frac{1}{3}x^3 - \frac{1}{2}x^2 - 6x - 4$
21. $f(x) = xe^x$
22. $f(x) = xe^{-x}$
23. $f(x) = x \ln x$

5

Basic concepts of integral calculus

26. Indefinite integrals of a function

To introduce the concept of indefinite integral we will consider the function $f(x) = 2x$. One idea that we have explored is connected with the derivative of $f(x) = 2x$, which we have seen to be $f'(x) = 2$. Another question, which takes us in the opposite direction, is this: What function has $f(x) = 2x$ as its derivative? That is, what function F has the property $F'(x) = 2x$? One answer to this question is

$$F(x) = x^2.$$

Another answer is

$$F(x) = x^2 + 1.$$

More generally,

$$F(x) = x^2 + C$$

where C is any constant, is a family of functions with $F'(x) = 2x$. All of these functions $F(x)$ are called indefinite integrals or antiderivatives of $f(x) = 2x$.

More generally, if f is a function, then any function F with the property that

$$F' = f$$

is called an *indefinite integral* or *antiderivative* of f. The notation

$$\int f(x)\, dx$$

read, "the integral of f of x, dx," is used to denote any one of the indefinite integrals of f. If f has an indefinite integral F to begin with, then it has many indefinite integrals, any one of which can be obtained from F by adding to it a suitable constant C. If F is an indefinite integral of f and $\int f(x)\,dx$ is any other indefinite integral of f, then

$$\int f(x)\,dx = F(x) + C$$

for some suitable constant C. To *integrate* a function $f(x)$ means to determine $F(x) = \int f(x)\,dx$ such that $F'(x) = f(x)$.

As further illustrations we have the following results.

$$\int 2\,dx = 2x + C \qquad \text{since} \qquad \frac{d(2x + C)}{dx} = 2.$$

$$\int 1\,dx = x + C \qquad \text{since} \qquad \frac{d(x + C)}{dx} = 1.$$

More generally, for any constant K, we have the following.

$$\int K\,dx = Kx + C \qquad \text{since} \qquad \frac{d(Kx + C)}{dx} = K.$$

$$\int x\,dx = \tfrac{1}{2}x^2 + C \qquad \text{since} \qquad \frac{d(\tfrac{1}{2}x^2 + C)}{dx} = x.$$

$$\int x^2\,dx = \tfrac{1}{3}x^3 + C \qquad \text{since} \qquad \frac{d(\tfrac{1}{3}x^3 + C)}{dx} = x^2.$$

$$\int x^3\,dx = \tfrac{1}{4}x^4 + C \qquad \text{since} \qquad \frac{d(\tfrac{1}{4}x^4 + C)}{dx} = x^3.$$

In general, if $n \neq -1$,

$$\int x^n\,dx = \frac{1}{n+1}x^{n+1} + C \qquad \text{since} \qquad \frac{d\left(\frac{1}{n+1}x^{n+1} + C\right)}{dx} = x^n.$$

Thus we have the following illustrations.

$$\int \frac{dx}{x^4} = \int x^{-4}\,dx = \frac{1}{-4+1}x^{-4+1} + C = -\frac{1}{3}x^{-3} + C$$

$$\int \sqrt{x}\,dx = \int x^{\frac{1}{2}}\,dx = \frac{1}{\frac{1}{2}+1}x^{\frac{1}{2}+1} + C = \frac{2}{3}x^{\frac{3}{2}} + C$$

$$\int \frac{dx}{\sqrt{x}} = \int x^{-\frac{1}{2}}\,dx = \frac{1}{-\frac{1}{2}+1}x^{-\frac{1}{2}+1} + C = 2x^{\frac{1}{2}} + C$$

For $n = -1$ and $x > 0$ we have

$$\int \frac{dx}{x} = \ln x + C \qquad \text{since} \qquad \frac{d(\ln x + C)}{dx} = \frac{1}{x}.$$

For $f(x) = e^x$ we have

$$\int e^x \, dx = e^x + C \qquad \text{since} \qquad \frac{d(e^x + C)}{dx} = e^x.$$

Two useful theorems for determining indefinite integrals are integral calculus counterparts of differentiation theorems. The counterpart of the sum theorem for derivatives—the derivative of a sum of two functions is the sum of their derivatives—is

$$\int [f(x) + g(x)] \, dx = \int f(x) \, dx + \int g(x) \, dx.$$

Thus to find an integral of a function which is a sum of two functions, find integrals of the component parts and add them. More generally, this theorem holds for a sum of two or more functions.

$$\int (x^4 + x^2 + 3) \, dx = \int x^4 \, dx + \int x^2 \, dx + \int 3 \, dx = \tfrac{1}{5}x^5 + \tfrac{1}{3}x^3 + 3x + C$$

$$\int (x^2 + e^x) \, dx = \int x^2 \, dx + \int e^x \, dx = \tfrac{1}{3}x^3 + e^x + C$$

The integral calculus counterpart of the theorem that states the derivative of a constant times a function equals the constant times the derivative of the function is

$$\int K \cdot f(x) \, dx = K \int f(x) \, dx.$$

Thus to find an integral of a constant times a function, find an integral of the function and multiply by the constant.

$$\int 5x^2 \, dx = 5 \int x^2 \, dx = 5(\tfrac{1}{3}x^3) + C = \tfrac{5}{3}x^3 + C$$

$$\int \left(4x^5 + \frac{3}{x}\right) dx = 4 \int x^5 \, dx + 3 \int \frac{dx}{x} = \frac{4}{6}x^6 + 3 \ln x + C$$

Exercises

Determine the following integrals and check your answers by differentiation.

1. $\int x^6 \, dx$

2. $\int x^{10} \, dx$

3. $\int x^{-5} \, dx$

4. $\int \frac{dx}{x^7}$

5. $\int x^{\frac{2}{3}} dx$

6. $\int x^{\frac{3}{4}} dx$

7. $\int 5x^{\frac{4}{3}} dx$

8. $\int 3x^{\frac{1}{5}} dx$

9. $\int 5x^{-\frac{3}{4}} dx$

10. $\int \frac{4}{x^{\frac{3}{2}}} dx$

11. $\int 4x^{\frac{1}{3}} dx$

12. $\int (x^2 + 1) dx$

13. $\int (x^3 + x^2 + 1) dx$

14. $\int (2x^4 + 3x + 1) dx$

15. $\int (3x^2 + 2x + 7) dx$

16. $\int \left(4\sqrt{x} - \frac{3}{x^3} \right) dx$

17. $\int \left(\frac{4}{\sqrt{x}} + 3x^2 \right) dx$

18. $\int (3x^{-\frac{2}{3}} - 4e^x) dx$

19. $\int \left(4x^3 - 3x - \frac{4}{x} \right) dx$

20. $\int \left(5x^6 - \frac{1}{x} \right) dx$

21. $\int \left(3e^x - 4 + \frac{1}{x^2} \right) dx$

22. $\int \left(\frac{1 + x}{x^2} \right) dx$

A change of variable stratagem

Many stratagems, some of considerable ingenuity, have been devised to determine indefinite integrals of functions. One such stratagem is a change of variable technique, of widespread applicability, which is the integral calculus counterpart of the chain rule of differential calculus. Suppose the integral $\int f(x)\, dx$ is being sought and the function $f(x)$ can be replaced by a simpler function of another variable u by introducing a suitable auxiliary function. Then the integral problem in terms of variable x can be stated as an integral problem in terms of variable u by use of the following theorem.

$$\int f(u) \cdot \frac{du}{dx} dx = \int f(u)\, du$$

where $u = u(x)$ is a function of x.

To illustrate, consider the problem of determining

$$F(x) = \int 2(1 + 2x)^3 dx.$$

We introduce the auxiliary function $u = 1 + 2x$; then $\dfrac{du}{dx} = 2$. Thus we obtain

$$\int \underbrace{2}_{\frac{du}{dx}} \underbrace{(1 + 2x)^3}_{u} dx = \int \underbrace{u^3}_{f(u)}\, du = \frac{1}{4} u^4 + C.$$

Replacing u by $1 + 2x$ yields

$$F(x) = \frac{1}{4}(1 + 2x)^4 + C$$

as the indefinite integrals of $f(x) = 2(1 + 2x)^3$. A specific indefinite integral is obtained by giving the constant C a specific value.

To verify our result we only need differentiate $F(x) = \frac{1}{4}(1 + 2x)^4 + C$ (with the aid of the chain rule) and observe that $F'(x) = f(x) = 2(1 + 2x)^3$.

EXAMPLE 1.   Determine $\int x \sqrt{3 + x^2}\, dx$.

Solution.   Let $u = 3 + x^2$ be our auxiliary function; then $\dfrac{du}{dx} = 2x$. Since there is an x term in $\int x \sqrt{3 + x^2}\, dx$, we need only multiply the integral by $\frac{1}{2}(2)$, which is 1, to obtain this term for $\dfrac{du}{dx}$. Thus we have the following results.

$$\int x \sqrt{3 + x^2}\, dx = \frac{1}{2} \int \underbrace{2x}_{\frac{du}{dx}} \underbrace{(3 + x^2)^{\frac{1}{2}}}_{u}\, dx$$

$$= \frac{1}{2} \int u^{\frac{1}{2}}\, du$$

$$= \frac{1}{2} \cdot \frac{2}{3} u^{\frac{3}{2}} + C$$

$$= \frac{1}{3} u^{\frac{3}{2}} + C$$

Replacing u by $3 + x^2$ yields

$$F(x) = \frac{1}{3}(3 + x^2)^{\frac{3}{2}} + C = \frac{1}{3}\sqrt{(3 + x^2)^3} + C$$

as the indefinite integrals of $f(x) = x\sqrt{3 + x^2}$.

EXAMPLE 2.   Determine $\int 1000\, e^{-0.1x}\, dx$.

Solution.   Let $u = -0.1x$ be our auxiliary function; then $\dfrac{du}{dx} = -0.1$.

$$\int 1000\, e^{-0.1x}\, dx = -\frac{1000}{0.1} \int \underbrace{(-0.1)}_{\frac{du}{dx}} \underbrace{e^{-0.1x}}_{u}\, dx = -\frac{1000}{0.1} \int e^u\, du = -\frac{1000}{0.1} e^u + C$$

Replacing u by $-0.1x$ yields

$$F(x) = -\frac{1000}{0.1} e^{-0.1x} + C = -10,000 e^{-0.1x} + C$$

as the indefinite integrals of $f(x) = 1000 e^{-0.1x}$.

EXAMPLE 3. Determine $\int 1000 e^{0.1(5-t)} dt$.

Solution. Let $u = 0.1(5 - t) = 0.5 - 0.1t$ be our auxiliary function; then $\frac{du}{dt} = -0.1$.

$$\int 1000 e^{0.1(5-t)} dt = -\frac{1000}{0.1} \int \underbrace{(-0.1)}_{\substack{| \\ \frac{du}{dt}}} \underbrace{e^{0.1(5-t)}}_{\substack{| \\ u}} dt$$

$$= -\frac{1000}{0.1} \int e^u du = -\frac{1000}{0.1} e^u + C$$

Replacing u by $0.1(5 - t)$ yields

$$F(t) = -10,000 e^{0.1(5-t)} + C$$

as the indefinite integrals of $f(t) = 1000 e^{0.1(5-t)}$.

EXAMPLE 4. Determine $\int \frac{10}{2x + 1} dx$.

Solution. Let $u = 2x + 1$ be our auxiliary function; then $\frac{du}{dx} = 2$.

$$\int \frac{10}{2x + 1} dx = 5 \int (2) \frac{1}{2x + 1} dx = 5 \int \frac{du}{u} = 5 \ln u + C$$

Replacing u by $2x + 1$ yields

$$F(x) = 5 \ln (2x + 1) + C$$

as the indefinite integrals of $f(x) = 10/(2x + 1)$.

Exercises

Determine the following indefinite integrals.

23. $\int 3(3x + 4)^2 dx$ 24. $\int 2x(x^2 + 1)^4 dx$

25. $\int 10x(5x^2 + 2)^3 dx$ 26. $\int x\sqrt{1 + x^2} dx$

27. $\int \dfrac{dx}{\sqrt{x-1}}$

28. $\int \dfrac{10x}{\sqrt{x^2+1}}\,dx$

29. $\int \dfrac{3x^2+1}{\sqrt{x^3+x}}\,dx$

30. $\int 2000\,e^{-0.5t}\,dt$

31. $\int 2000\,e^{-0.1t}\,dt$

32. $\int \dfrac{2x}{x^2-1}\,dx$

33. $\int \dfrac{3x^2+1}{x^3+x}\,dx$

34. $\int x^3\sqrt{10+x^4}\,dx$

35. $\int A\,e^{-rt}\,dt$

36. $\int 2000\,e^{0.1(10-t)}\,dt$

37. $\int 2000\,e^{0.05(10-t)}\,dt$

38. $\int A\,e^{r(x-t)}\,dt$

39. $\int \dfrac{dx}{x+4}$

40. $\int \dfrac{x}{x^2+1}\,dx$

41. $\int \dfrac{2x-1}{x^2-x}\,dx$

42. $\int xe^{-x^2}\,dx$

43. $\int e^{(2x+5)}\,dx$

44. $\int 3x\sqrt{5-x^2}\,dx$

45. $\int \sqrt{3-7x}\,dx$

46. $\int x^2e^{x^3}\,dx$

Integration by parts

The product theorem of differential calculus states that if $h(x) = f(x) \cdot g(x)$ and $f'(x)$ and $g'(x)$ exist, then

$$\frac{df(x) \cdot g(x)}{dx} = f(x) \cdot g'(x) + g(x) \cdot f'(x). \tag{1}$$

The integral calculus counterpart of this product theorem, called *integration by parts,* is obtained by integrating each member of equation (1).

$$f(x) \cdot g(x) = \int f(x) \cdot g'(x)\,dx + \int g(x) \cdot f'(x)\,dx$$

from which we obtain

$$\int f(x) \cdot g'(x)\,dx = f(x) \cdot g(x) - \int g(x) \cdot f'(x)\,dx. \tag{2}$$

To profitably employ the integration-by-parts stratagem, we must be able to choose $f(x)$ and $g'(x)$ so that $g'(x)$ is readily integrated and the integral $\int g(x) \cdot f'(x)\,dx$ is easier to handle than the given integral.

EXAMPLE 5. Determine $\int xe^x\,dx$.

Solution. Choose $f(x)$ and $g'(x)$ as follows.

$$f(x) = x \quad \text{and} \quad g'(x) = e^x$$

Then $f'(x)$ and $g(x)$ are

$$f'(x) = 1 \quad \text{and} \quad g(x) = \int e^x \, dx = e^x.$$

From the integration-by-parts formula we obtain the following results. Note here that we wait until we have the final expression for the integral before adding the constant C.

$$\int xe^x \, dx = xe^x - \int e^x \, dx = xe^x - e^x + C$$

EXAMPLE 6. Determine $\int (2200 - 100t) \, e^{-0.1t} \, dt$.

Solution. Let $f(t) = 2200 - 100t$ and $g'(t) = e^{-0.1t}$. Then

$$f'(t) = -100 \quad \text{and} \quad g(t) = \int e^{-0.1t} \, dt = -\frac{1}{0.1} e^{-0.1t}.$$

From the integration-by-parts formula we obtain the following results.

$$\int (2200 - 100t) \, e^{-0.1t} \, dt = -\frac{(2200 - 100t)}{0.1} e^{-0.1t} - \int 1000 \, e^{-0.1t} \, dt$$

$$= -\frac{(2200 - 100t)}{0.1} e^{-0.1t} + \frac{1000}{0.1} e^{-0.1t} + C$$

$$= -22{,}000 \, e^{-0.1t} + 1000t \, e^{-0.1t} + 10{,}000 \, e^{-0.1t} + C$$

$$= -12{,}000 \, e^{-0.1t} + 1000t \, e^{-0.1t} + C$$

Exercises

Determine the following integrals.

47. $\int xe^{-x} \, dx$

48. $\int xe^{3x} \, dx$

49. $\int (1800 - 50t) \, e^{-0.1t} \, dt$

50. $\int (1000 - 20t) \, e^{-0.05t} \, dt$

51. $\int xe^{-2x} \, dx$

52. $\int (100 + 3t) \, e^{0.1(5-t)} \, dt$

53. $\int (500 - 10t) \, e^{-0.08t} \, dt$

54. $\int \ln x \, dx$

55. $\int \frac{x^3}{\sqrt{1 + x^2}} \, dx$

56. $\int x \ln x \, dx$

57. $\int x^2 \ln x \, dx$

58. $\int x^2 e^{-x} \, dx$

Problems involving indefinite integrals

EXAMPLE 7. Find the curves whose tangent line at any point (x, y) has slope $2x$. Determine the curve which contains the point $(1, -1)$.

Solution. Suppose $y = f(x)$ has as its graph one of the curves we seek; then $\dfrac{dy}{dx} = 2x$. By integrating we obtain the following result.

$$y = \int 2x \, dx = x^2 + C$$

$y = x^2 + C$ describes a family of curves called parabolas. A specific member of this family is determined by giving C a specific value. For $C = 0$ we obtain

$$y = x^2.$$

For $C = 1$ we obtain

$$y = x^2 + 1.$$

For $C = -2$ we obtain

$$y = x^2 - 2.$$

All of the members of this family of parabolas have slope $2x$ at any point (x, y).

To determine the member of this family of parabolas which contains the point $(1, -1)$, we must determine C so that $y = x^2 + C$ is satisfied by $(1, -1)$. To determine C, substitute 1 for x and -1 for y in $y = x^2 + C$ and solve for C.

$$-1 = (1)^2 + C$$
$$C = -2$$

Thus $y = x^2 - 2$ has slope $2x$ at (x, y) and contains the point $(1, -1)$. In Figure 1 the members of the family of parabolas $y = x^2 + C$ when C is -2, 0, and 1 are shown.

EXAMPLE 8. The marginal cost function of a sugar refinery is $c'(x) = \frac{2}{10}x + 2$, where x is output in tons per week. If the sugar refinery incurs a fixed cost of \$200 per week, determine the total cost function $c(x)$.

Solution. To determine $c(x)$ we integrate $c'(x) = \frac{2}{10}x + 2$.

$$c(x) = \int (\tfrac{2}{10}x + 2) \, dx = \tfrac{1}{10}x^2 + 2x + C$$

Since the constant factor in the total cost function represents fixed costs, we have $C = 200$. Thus the total cost function is

$$c(x) = \tfrac{1}{10}x^2 + 2x + 200.$$

EXAMPLE 9. Income from a certain investment is obtained continuously over time at a rate of $f(t) = 1000t$ dollars per year, t years from the present ($t = 0$). If the income at time

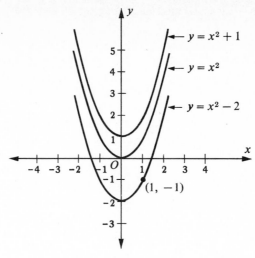

Figure 1

$t = 0$ is \$2000, determine the income accumulation function $R(t)$ and the income accumulated after two years.

Solution.   Since, by definition, the rate of income accumulation $f(t)$ is the derivative of the income accumulation function $R(t)$, we must integrate $f(t)$ to obtain $R(t)$. Thus for the situation given we obtain

$$R(t) = \int 1000t \, dt = 500t^2 + C.$$

To determine C we use the condition that $R(0) = 2000$.

$$R(0) = 500(0)^2 + C$$
$$2000 = C$$

Thus

$$R(t) = 500t^2 + 2000.$$

The income accumulated after two years is

$$R(2) = 500(2)^2 + 2000 = \$4000.$$

Exercises

59. Find the curves whose tangent line at any point (x, y) has slope $3x^2 + 2$. Determine the curve which contains the point $(2, 15)$.
60. Find the equation of the curve whose tangent line at any point (x, y) has slope $2x^2 + 3x + 1$, and the curve contains the point $(-1, 5)$.
61. Find the equation of the curve whose tangent line at any point (x, y), $x \neq 0$, has slope $-(1/x^2)$, and the curve contains $(1, -3)$.

62. The marginal cost function of a coffee producer is

$$c'(x) = \frac{100}{2\sqrt{100x + 2000}}$$

for an output of x tons of coffee per day. If fixed costs are \$150 per day, determine the coffee producer's total cost function $c(x)$.

63. The marginal cost function of a rare-metal producer is $c'(x) = 2000\, e^{2x}$ for an output of x tons per week. If fixed costs are \$1000 per week, determine the rare-metal producer's total cost function $c(x)$.

64. The marginal revenue function of a sugar refinery is $R'(x) = 100 - 10x$ for an output of x tons a day with a total revenue of $R(x)$ dollars. Determine the total revenue function $R(x)$.

65. Income from a certain investment is obtained continuously over time at a rate of $f(t) = 2000\sqrt{t}$ dollars per year, t years from the present $(t = 0)$. If the income at time $t = 0$ is \$1000, determine the income accumulation function $R(t)$ and the income accumulated after three years.

66. The instantaneous velocity at time t of a particle moving in a straight line path is $v(t) = 2t + 2$. If the distance moved by the particle at time $t = 2$ seconds is eight feet, determine the time-distance function which describes the motion of the particle.

67. The marginal revenue function of a cement manufacturer is $R'(x) = 200 - 3x$ for an output of x tons a day with a total revenue of $R(x)$ dollars. Determine the total revenue function and the total revenue for an output of six tons a day.

68. Income from a certain investment is generated continuously over time at the rate of $f(t) = 1000\, e^t - 100t$ dollars per year, t years from the present $(t = 0)$. If the income at time $t = 0$ is \$19,000, determine the income accumulation function and the income accumulated after two years $(e^2 = 7.3891)$.

69. Find the equation of the curve whose tangent line at any point (x, y) has slope $x/(x^2 + 1)$, and the curve contains point $(0, 5)$.

27. Situations with a common structure

To prepare the way for our discussion of the definite integral of a function, we first discuss two of many different situations which have this structure in common. The concept of definite integral will emerge in the course of distilling out the common features of the situations. We will study this structure in its own right without reference to the various situations which exhibit it. Then we will return to some situations with an integral structure and apply to them whatever results we have obtained.

In our discussions we will have occasion to consider sums of quantities, and it will be useful to have a suitable notation for these sums. The Greek letter Σ (sigma) is used in mathematics to denote a sum. Specifically, consider the sum

$$2^1 + 2^2 + 2^3 + 2^4 + 2^5 + 2^6 + 2^7 + 2^8.$$

In terms of the Σ notation, this sum would be denoted by

$$\sum_{i=1}^{i=8} 2^i.$$

We introduce a variable—here i is used—to represent the quantity which is changing from term to term in the sum (the exponent in this case). Below the Σ sign we indicate the initial value of the variable ($i = 1$), and above the Σ sign we specify the terminal value of the variable ($i = 8$). Thus the sum

$$x_1 + x_2 + x_3 + x_4 + x_5 + x_6$$

is denoted by

$$\sum_{i=1}^{i=6} x_i.$$

The sum

$$x_0(x_1 - x_0) + x_1(x_2 - x_1) + x_2(x_3 - x_2) + x_3(x_4 - x_3)$$

is denoted by

$$\sum_{i=1}^{i=4} x_{i-1}(x_i - x_{i-1}) \quad \text{or} \quad \sum_{i=0}^{i=3} x_i(x_{i+1} - x_i).$$

The sum

$$f(m_1) \cdot (x_1 - x_0) + f(m_2) \cdot (x_2 - x_1) + \cdots + f(m_n) \cdot (x_n - x_{n-1})$$

is denoted by

$$\sum_{i=1}^{i=n} f(m_i) \cdot (x_i - x_{i-1}).$$

The concepts of least upper bound (LUB) and greatest lower bound (GLB) of a set of numbers also play an important role in our discussions and it would therefore be in order to discuss them at this point.

Let B denote a set of numbers. A number k is said to be an *upper bound* of set B if every number in B is less than or equal to k. The *least upper bound* of B (LUB of B) is a number I such that

1. I is an upper bound of set B;
2. no number less than I is an upper bound of B, that is, I is the smallest of the upper bounds of B.

If $B = \{\frac{1}{2}, 3, \sqrt{10}, 15, 27\}$, then 27 is an upper bound of B and any number greater than 27 is an upper bound of B. $27 = $ LUB of B. If B is the set of numbers between

0 and 2, then 2 is an upper bound of B, as is any number greater than 2. LUB of $B = 2$.

A number c is said to be a *lower bound* of set B if every number in B is greater than or equal to c. The *greatest lower bound* of B (GLB of B) is a number J such that

1. J is a lower bound of set B;
2. no number greater than J is a lower bound of B, that is, J is the largest of the lower bounds of B.

If $B = \{\frac{1}{2}, 3, \sqrt{10}, 15, 27\}$, then $\frac{1}{2}$ is a lower bound of B, as is any number less than $\frac{1}{2}$. GLB of $B = \frac{1}{2}$. If B is the set of numbers between 0 and 2, then 0 is a lower bound of B, as is any number less than 0. GLB of $B = 0$.

Exercises

Express the following sums in terms of the Σ notation.

70. $(1 - \sqrt{1}) + (1 - \sqrt{2}) + (1 - \sqrt{3}) + (1 - \sqrt{4})$
71. $(1 + 2^1) + (2 + 2^2) + (3 + 2^3) + (4 + 2^4)$
72. $(4 - \frac{1}{2}) + (5 - \frac{2}{3}) + (6 - \frac{3}{4}) + (7 - \frac{4}{5})$
73. $1^2(x_1 - x_0) + 2^2(x_2 - x_1) + 3^2(x_3 - x_2) + \cdots + n^2(x_n - x_{n-1})$
74. $x_1 x_2 + x_2 x_3 + x_3 x_4 + \cdots + x_{n-1} x_n$
75. $x_1(x_2 + x_4) + x_2(x_3 + x_5) + x_3(x_4 + x_6) + x_4(x_5 + x_7)$
76. In a discussion of the set of numbers between 2 and 5 it was said that the least upper bound of this set is 4.99 and the greatest lower bound is 1.99. Do you agree? Explain.
77. Determine the least upper bound of the set of numbers between 1 and 9. What is the basis for your conclusion?
78. Determine the greatest lower bound of the set of numbers between 1 and 9. What is the basis for your conclusion?
79. Determine the least upper bound and greatest lower bound of the set $\left\{1, \frac{1}{2}, \frac{1}{3}, \frac{1}{4}, \ldots, \frac{1}{n}, \ldots\right\}$. What is the basis for your conclusions?

An area problem

In geometry, the area $A(R)$ of a rectangle R with height h and base of length k is defined as the product of h and k. $A(R)$ serves as a numerical measure of the amount of space occupied by R. The properties of $A(R)$ include the following.

1. If r is a rectangle which is contained in R, then $A(r) \leq A(R)$.
2. If R is partitioned into nonoverlapping subrectangles R_1 and R_2 (see Figure 2), then $A(R) = A(R_1) + A(R_2)$.

In geometry it is shown that the concept of area can be extended from rectangles to other polygons. The aforementioned properties of area are preserved in this extension.

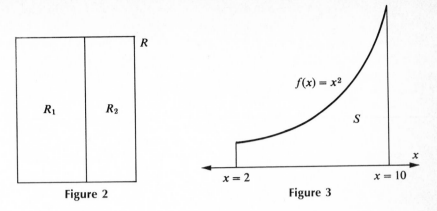

Figure 2

Figure 3

1. If D and E are polygons and D is contained in E, then $A(D) \leq A(E)$.
2. If D is partitioned into several nonoverlapping subpolygons, then $A(D)$ is equal to the sum of the areas of these nonoverlapping subpolygons.

With this background in mind let us consider a related problem. We want to extend the concept of area to regions which are not bounded by line segments and, in addition, we want to preserve the fundamental properties for the area concept, such as properties 1 and 2, defined for rectangles and other polygons. Let us begin by considering a very specific situation. Consider the region S bounded below by the x-axis, on the sides by the lines $x = 2$ and $x = 10$, and above by the graph of the function $f(x) = x^2$ (see Figure 3). Now we want to define the area $A(S)$.

To begin, let m_1 and M_1 denote values in the interval $[2, 10]^*$ at which $f(x) = x^2$ takes on its absolute minimum value and absolute maximum value, respectively. Clearly, from Figure 4, $m_1 = 2$ and $M_1 = 10$, so that the minimum value and maximum value of $f(x) = x^2$ in $[2, 10]$ are 4 and 100, respectively. Let r_1 denote

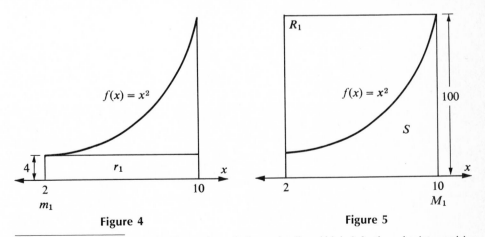

Figure 4

Figure 5

---

*As you recall, the notation $[a, b]$ describes an interval of a number line which includes the end points a and b as well as all points between a and b.

an inner rectangle with base [2, 10] and height $f(m_1) = f(2) = 4$. r_1 is completely contained in region S (see Figure 4). Let R_1 denote an outer rectangle with base [2, 10] and height $f(M_1) = f(10) = 100$. Region S is completely contained in R_1 (see Figure 5). However $A(S)$ is defined, if it is to be consistent with our geometric experience, it should neither be less than $A(r_1)$ nor exceed $A(R_1)$. In short, $A(S)$ should satisfy the condition

$$A(r_1) \le A(S) \le A(R_1).$$

This condition, by definition of area for rectangles, becomes

$$f(m_1) \cdot (10 - 2) \le A(S) \le f(M_1) \cdot (10 - 2)$$
$$32 \le A(S) \le 800.$$

We thus obtain 32 as a lower bound and 800 as an upper bound for $A(S)$.

To narrow the gap, we partition S into two subregions, S_1 and S_2, and repeat the argument just given for each of them. Take a value between 2 and 10, 5 for example. 5 partitions the interval [2, 10] into the two subintervals, [2, 5] and [5, 10], and thus partitions S into the two subregions S_1 and S_2 with [2, 5] and [5, 10] as their respective bases (see Figure 6).

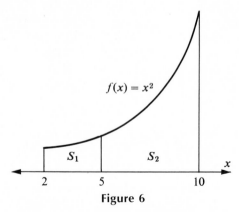

Figure 6

Let m_1 and m_2 denote values in [2, 5] and [5, 10] at which $f(x) = x^2$ takes on its absolute minimum value in these respective subintervals ($m_1 = 2$, $m_2 = 5$). Let M_1 and M_2 denote values in [2, 5] and [5, 10] at which $f(x) = x^2$ takes on its absolute maximum value in these respective subintervals ($M_1 = 5$, $M_2 = 10$).

As before, this introduction of minimum and maximum values leads to inner rectangles r_1 and r_2 which are contained in S_1 and S_2 (see Figure 7) and outer rectangles R_1 and R_2 which contain S_1 and S_2 (see Figure 8). If our definition of area is to be in accord with out intuitive conception of area, then $A(S_1)$ and $A(S_2)$, the areas of S_1 and S_2 to be defined, should satisfy these conditions.

$$A(r_1) \le A(S_1) \le A(R_1)$$
$$A(r_2) \le A(S_2) \le A(R_2)$$

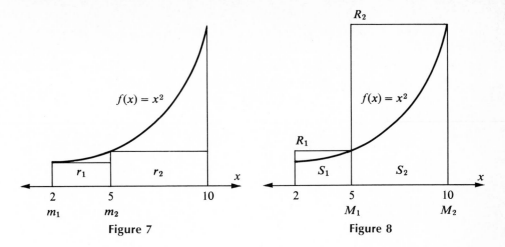

Figure 7   **Figure 8**

In this case then

$$f(m_1) \cdot (5 - 2) \le A(S_1) \le f(M_1) \cdot (5 - 2)$$
$$f(m_2) \cdot (10 - 5) \le A(S_2) \le f(M_2) \cdot (10 - 5)$$

from which we obtain

$$12 \le A(S_1) \le 75$$
$$125 \le A(S_2) \le 500.$$

Addition yields

$$137 \le A(S_1) + A(S_2) \le 575.$$

Since S_1 and S_2 were obtained by partitioning S into two nonoverlapping pieces, it is reasonable to impose on $A(S)$ the condition

$$A(S) = A(S_1) + A(S_2).$$

We thereby obtain

$$137 \le A(S) \le 575.$$

We have thus raised our lower bound from 32 to 137 and lowered our upper bound from 800 to 575.

If we partition S into three subregions, S_1, S_2, and S_3, by choosing two values in [2, 10], say 4 and 5 (see Figures 9, 10, and 11), then by analogous arguments we obtain

$$A(r_1) \le A(S_1) \le A(R_1)$$
$$A(r_2) \le A(S_2) \le A(R_2)$$
$$A(r_3) \le A(S_3) \le A(R_3).$$

Sec 27   Situations with a common structure

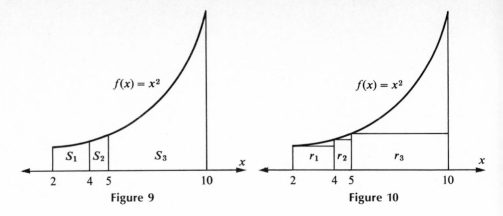

Figure 9   **Figure 10**

These conditions yield

$$8 \leq A(S_1) \leq 32$$
$$16 \leq A(S_2) \leq 25$$
$$125 \leq A(S_3) \leq 500$$

from which we obtain

$$149 \leq A(S) \leq 557.$$

We have thus raised the lower bound to 149 and lowered the upper bound to 557.

Partitioning S into four subregions, S_1, S_2, S_3, and S_4, by means of the partition points 2, 4, 5, 8, and 10, yields, by use of analogous arguments,

$$227 \leq A(S) \leq 449$$

thereby raising the lower bound to 227 and lowering the upper bound to 449.

The idea illustrated by these specific partitionings of S into subregions is as basic as it is simple. Lower and upper bounds for $A(S)$ can be obtained by partitioning

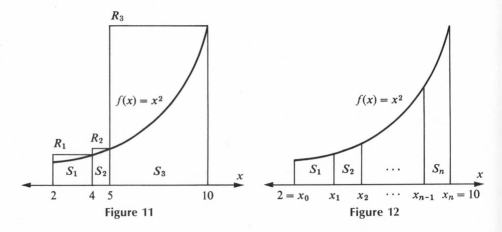

Figure 11   **Figure 12**

S into subregions and constructing inner and outer rectangles as indicated for each subregion. The sum of the areas of the inner rectangles serve as a lower bound for $A(S)$ and the sum of the areas of the outer rectangles serve as an upper bound for $A(S)$.

Stated in more precise language, S can be partitioned into n subregions, S_1, $S_2, \ldots, S_n$, by choosing a sequence of partition points $2 = x_0, x_1, x_2, \ldots, x_{n-1}$, $x_n = 10$ (see Figure 12). S_1 has base $[x_0, x_1]$, S_2 has base $[x_1, x_2]$, and so on. The inner rectangles $r_1, r_2, \ldots, r_n$, which are contained in $S_1, S_2, \ldots, S_n$, are shown in Figure 13, and the outer rectangles $R_1, R_2, \ldots, R_n$, which contain $S_1, S_2, \ldots,$ S_n, are shown in Figure 14.

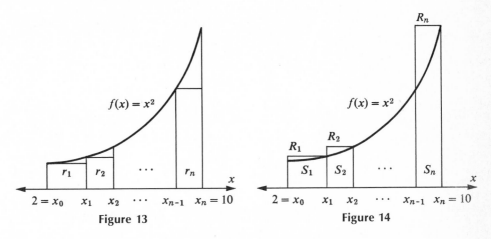

Figure 13   **Figure 14**

Our intuitive conception of area leads us to impose these conditions.

$$A(r_1) \leq A(S_1) \leq A(R_1)$$
$$A(r_2) \leq A(S_2) \leq A(R_2)$$
$$\vdots \qquad \vdots \qquad \vdots$$
$$A(r_n) \leq A(S_n) \leq A(R_n)$$

Addition yields

$$\underbrace{A(r_1) + \cdots + A(r_n)}_{\substack{\text{sum of the areas of} \\ \text{the inner rectangles}}} \leq A(S_1) + \cdots + A(S_n) \leq \underbrace{A(R_1) + \cdots + A(R_n)}_{\substack{\text{sum of the areas of} \\ \text{the outer rectangles}}}.$$

Since, to make intuitive sense, $A(S)$ should have the property

$$A(S) = A(S_1) + \cdots + A(S_n)$$

we obtain

$$A(r_1) + \cdots + A(r_n) \leq A(S) \leq A(R_1) + \cdots + A(R_n).$$

To obtain expressions for $A(r_1), \ldots, A(r_n)$ and $A(R_1), \ldots, A(R_n)$, let $m_1, m_2, \ldots, m_n$ denote values in $[x_0, x_1], [x_1, x_2], \ldots, [x_{n-1}, x_n]$ at which $f(x) = x^2$ takes on its absolute minimum value in these respective subintervals ($m_1 = x_0$, $m_2 = x_1$, ..., $m_n = x_{n-1}$ in this case). Let $M_1, M_2, \ldots, M_n$ denote values in $[x_0, x_1], [x_1, x_2], \ldots, [x_{n-1}, x_n]$ at which $f(x) = x^2$ takes on its absolute maximum value in these respective subintervals ($M_1 = x_1$, $M_2 = x_1$, ..., $M_n = x_n$ in this case). Then

$$A(r_1) = f(m_1) \cdot (x_1 - x_0) \qquad A(R_1) = f(M_1) \cdot (x_1 - x_0)$$
$$A(r_2) = f(m_2) \cdot (x_2 - x_1) \qquad A(R_2) = f(M_2) \cdot (x_2 - x_1)$$
$$\vdots \qquad\qquad\qquad \vdots$$
$$A(r_n) = f(m_n) \cdot (x_n - x_{n-1}) \qquad A(R_n) = f(M_n) \cdot (x_n - x_{n-1}).$$

We thus obtain

$$\underbrace{f(m_1) \cdot (x_1 - x_0) + \cdots + f(m_n) \cdot (x_n - x_{n-1})}_{\substack{\text{sum of the areas of the inner} \\ \text{rectangles}}} \leq A(S)$$

$$\leq \underbrace{f(M_1) \cdot (x_1 - x_0) + \cdots + f(M_n) \cdot (x_n - x_{n-1}).}_{\substack{\text{sum of the areas of the outer} \\ \text{rectangles}}}$$

This condition is expressed in terms of Σ notation by

$$\underbrace{\sum_{i=1}^{i=n} f(m_i) \cdot (x_i - x_{i-1})}_{\substack{\text{sum of the areas} \\ \text{of the inner} \\ \text{rectangles}}} \leq A(S) \leq \underbrace{\sum_{i=1}^{i=n} f(M_i) \cdot (x_i - x_{i-1}).}_{\substack{\text{sum of the areas} \\ \text{of the outer} \\ \text{rectangles}}}$$

Such bounds are obtained for $A(S)$ for every sequence of partition points $2 = x_0$, $x_1, \ldots, x_{n-1}, x_n = 10$ used.

We have seen that by partitioning S into more and smaller subregions, we narrow the gap between the lower and upper bounds obtained for $A(S)$. The problem is to close the gap entirely. The concepts of least upper bound and greatest lower bound of a set offer us a means of doing this.

Since every lower-bound sum $\Sigma f(m_i) \cdot (x_i - x_{i-1})$ is less than or equal to $A(S)$, the least upper bound of the set of all lower-bound sums, $\{\Sigma f(m_i) \cdot (x_i - x_{i-1})\}$, which we will denote by I, is less than or equal to $A(S)$.

$$I = \text{LUB of } \left\{ \sum_{i=1}^{i=n} f(m_i) \cdot (x_i - x_{i-1}) \right\}$$

$$I \leq A(S)$$

Since every upper-bound sum $\Sigma f(M_i) \cdot (x_i - x_{i-1})$ is greater than or equal to $A(S)$, then the greatest lower bound of the set of all upper-bound sums, $\{\Sigma f(M_i) \cdot (x_i - x_{i-1})\}$, which we denote by J, is greater than or equal to $A(S)$.

$$J = \text{GLB of } \left\{ \sum_{i=1}^{i=n} f(M_i) \cdot (x_i - x_{i-1}) \right\}$$

$$A(S) \leq J$$

We thus have

$$I \leq A(S) \leq J.$$

If $I = J$, then we define $A(S)$ to be their common value.

While this definition is certainly reasonable, it raises some important and difficult questions. Under what conditions does $I = J$? How can the common value of I and J be determined? We will give answers to these questions in the next section.

More generally, consider a region S bounded below by the x-axis, on the sides by the lines $x = a$ and $x = b$, and above by the graph of a function $y = f(x)$ which is continuous for all x satisfying $a \leq x \leq b$ (see Figure 15). Partition S into subregions $S_1, S_2, \ldots, S_n$ by choosing a sequence of points $a = x_0, x_1, x_2, \ldots, x_{n-1}$, $x_n = b$ (see Figure 16). Each subregion is bounded by its inner and outer rectangles (see Figures 17 and 18).

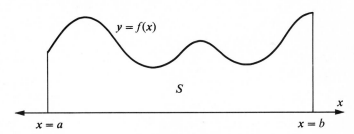

Figure 15

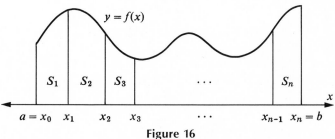

Figure 16

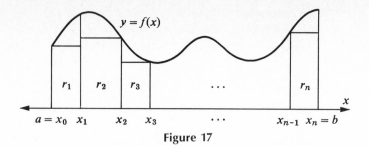

Figure 17

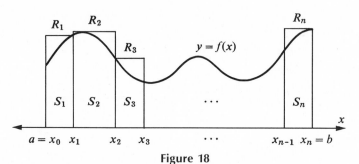

Figure 18

Our intuitive idea of area leads us to impose these conditions.

$$A(r_1) \leq A(S_1) \leq A(R_1)$$
$$A(r_2) \leq A(S_2) \leq A(R_2)$$
$$\vdots \qquad \vdots \qquad \vdots$$
$$A(r_n) \leq A(S_n) \leq A(R_n)$$

By adding we obtain

$$A(r_1) + \cdots + A(r_n) \leq A(S_1) + \cdots + A(S_n) \leq A(R_1) + \cdots + A(R_n).$$

Since $A(S)$ should be equal to the sum of the areas of its component subregions, that is,

$$A(S) = A(S_1) + \cdots + A(S_n)$$

we have

$$A(r_1) + \cdots + A(r_n) \leq A(S) \leq A(R_1) + \cdots + A(R_n).$$

Now we also know the following information.

$$A(r_1) = f(m_1) \cdot (x_1 - x_0) \qquad A(R_1) = f(M_1) \cdot (x_1 - x_0)$$
$$A(r_2) = f(m_2) \cdot (x_2 - x_1) \qquad A(R_2) = f(M_2) \cdot (x_2 - x_1)$$
$$\vdots \qquad\qquad\qquad \vdots$$
$$A(r_n) = f(m_n) \cdot (x_n - x_{n-1}) \qquad A(R_n) = f(M_n) \cdot (x_n - x_{n-1})$$

Here $m_1, m_2, \ldots, m_n$ are points in $[x_0, x_1], [x_1, x_2], \ldots, [x_{n-1}, x_n]$ at which $y = f(x)$ takes on its absolute minimum value in these respective subintervals. $M_1, M_2, \ldots,$

M_n are values in $[x_0, x_1]$, $[x_1, x_2]$, ..., $[x_{n-1}, x_n]$ at which $y = f(x)$ takes on its absolute maximum value in these respective subintervals.

Thus we have

$$\underbrace{\sum_{i=1}^{i=n} f(m_i) \cdot (x_i - x_{i-1})}_{\substack{\text{sum of the areas of} \\ \text{the inner rectangles}}} \leq A(s) \leq \underbrace{\sum_{i=1}^{i=n} f(M_i) \cdot (x_i - x_{i-1})}_{\substack{\text{sum of the areas of} \\ \text{the outer rectangles}}}.$$

Such bounds are obtained for every sequence of partition points $a = x_0$, x_1, x_2, ..., x_{n-1}, $x_n = b$.

If we let

$$I = \text{LUB of } \left\{ \sum_{i=1}^{i=n} f(m_i) \cdot (x_i - x_{i-1}) \right\}$$

and

$$J = \text{GLB of } \left\{ \sum_{i=1}^{i=n} f(M_i) \cdot (x_i - x_{i-1}) \right\}$$

then we obtain

$$I \leq A(S) \leq J.$$

If $I = J$, then $A(S)$ is defined as their common value.

Exercises

80. Consider the region S bounded below by the x-axis, on the left and right by the lines $x = 2$ and $x = 8$, and above by the graph of $f(x) = 2x^2 + 1$.
    a. Determine lower and upper bounds for the area $A(S)$ of S on the basis of the sequence of values $\langle 2, 4, 8 \rangle$ which partitions S into two subregions.
    b. Determine lower and upper bounds for $A(S)$ on the basis of the sequence $\langle 2, 3, 4, 8 \rangle$ which partitions S into three subregions.
    c. Determine lower and upper bounds for $A(S)$ on the basis of the sequence $\langle 2, 3, 4, 6, 8 \rangle$ which partitions S into four subregions.
81. Consider the region S bounded below by the x-axis, on the left and right by the lines $x = -2$ and $x = 3$, and above by the graph of $f(x) = x^2 + 2x + 2$.
    a. Determine lower and upper bounds for $A(S)$ on the basis of the sequence $\langle -2, -1, 3 \rangle$.
    b. Determine lower and upper bounds for $A(S)$ on the basis of the sequence $\langle -2, -1, 0, 3 \rangle$.
    c. Determine lower and upper bounds for $A(S)$ on the basis of the sequence $\langle -2, -1, 0, 1, 3 \rangle$.

A work problem

If a constant force moves an object a certain distance along a straight line path, then the product of the magnitude of the force times the distance moved is called the *work done* by the force in moving the object. Thus if 20 pounds is used to move an object 10 feet, the work done in moving the object is $20(10) = 200$ foot-pounds.

The work done by a constant force serves as a quantitative measure of the energy expended in moving an object a given distance. But suppose the force is not constant in magnitude. What would be an appropriate way to extend the concept of work to such a situation? This is the problem we now consider. To be specific, suppose that an object is moved six feet along a horizontal line by a variable force whose magnitude x feet from the initial position of the object is $f(x) = 2x + 3$ pounds (see Figure 19).

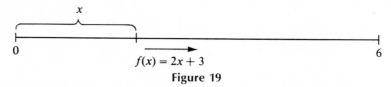

Figure 19

The problem, then, is to define a concept of work done by this force whose magnitude varies from point to point in moving the object six feet. However the work done by this force is defined, it should not be less than $3 \cdot 6 = 18$ foot-pounds since the minimum value of the force in the interval of motion is $f(0) = 3$ pounds and the total distance moved by the object is 6 feet. Also, the work done should not exceed $15 \cdot 6 = 90$ foot-pounds since the maximum value of the force in the interval of motion is $f(6) = 15$ pounds. Thus if we let W denote the work done by the force, we have

$$18 \le W \le 90$$

as the bounds for the work done.

Now suppose we regard the work done W as being done in two stages. In the first stage the object is moved, let us say, two feet, from 0 to 2, and in the second stage the object is moved four feet, from 2 to 6 (see Figure 20).

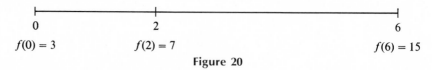

Figure 20

In the first stage the work done, call it W_1, should not be less than $3 \cdot 2 = 6$ foot-pounds (the minimum value of the force in this interval is $f(0) = 3$ pounds and the distance moved is 2 feet). Also, W_1 should not exceed $7 \cdot 2 = 14$ foot-pounds (the maximum value of the force in this interval is $f(2) = 7$ pounds and the distance moved is 2 feet). We thus have

$$6 \le W_1 \le 14.$$

In the second stage the work done, call it W_2, should not be less than $7 \cdot 4 = 28$ foot-pounds (the minimum value of the force in this interval is $f(2) = 7$ pounds and the distance moved is 4 feet). Also, W_2 should not exceed $15 \cdot 4 = 60$ foot-pounds (the maximum value of the force in this interval is $f(6) = 15$ pounds and the distance moved is 4 feet). We thus have

$$28 \leq W_2 \leq 60.$$

One of the properties of the work concept for a force of constant magnitude is that if W is the work done over an interval I, and W_1, W_2, etc., represent the work done over subintervals into which I has been partitioned, then

$$W = W_1 + W_2 + \cdots.$$

That is, the work done over an interval I equals the sum of the work done over the subintervals into which I has been partitioned.

If the concept of work for a variable force which we are seeking to define is to be modeled after the concept of work for a constant force, then it should share its most important properties. We are thus led to impose the condition that the work done by a variable force over an interval I be equal to the sum of the work done over subintervals into which I has been partitioned. For the situation under consideration, this leads to the condition

$$W = W_1 + W_2.$$

Since

$$6 \leq W_1 \leq 14$$
$$28 \leq W_2 \leq 60$$

addition yields

$$6 + 28 \leq W_1 + W_2 \leq 14 + 60$$

from which we obtain

$$34 \leq W \leq 74$$

as the bounds for W.

Suppose we regard the work done W as being done in three stages, where, let us say, in the first stage the object is moved two feet, from 0 to 2, in the second stage it is moved three feet, from 2 to 5, and in the third stage it is moved one foot, from 5 to 6 (see Figure 21). Then we obtain

$$6 \leq W_1 \leq 14$$
$$21 \leq W_2 \leq 39$$
$$13 \leq W_3 \leq 15$$

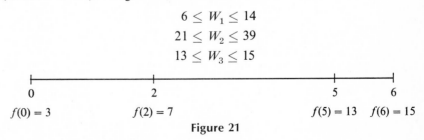

Figure 21

as the bounds for W_1, W_2, and W_3, the work done in each subinterval. By adding and using the condition

$$W = W_1 + W_2 + W_3$$

we obtain

$$40 \leq W \leq 68$$

as the bounds for W.

We thus see the same idea here in our work problem as in our area problem. A lower bound for W is obtained by partitioning the interval $[0, 6]$ into subintervals, forming the product of the minimum value of the force function in each subinterval times the length of the subinterval, then adding. An upper bound is obtained by forming the product of the maximum value of the force function in each subinterval times the length of the subinterval, then adding.

More generally, suppose we partition the interval $[0, 6]$ into n subintervals, $[x_0, x_1]$, $[x_1, x_2]$, ..., $[x_{n-1}, x_n]$, by choosing a sequence of partition points $0 = x_0, x_1, x_2$, ..., $x_{n-1}, x_n = 6$ (see Figure 22). Then suppose we regard the work done W as

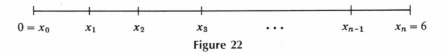

$$0 = x_0 \qquad x_1 \qquad x_2 \qquad x_3 \qquad \cdots \qquad x_{n-1} \qquad x_n = 6$$

Figure 22

being done in n stages, where in the first stage the object is moved from $0 = x_0$ to x_1, in the second stage it is moved from x_1 to x_2, in the third stage it is moved from x_2 to x_3, and so on. As bounds for W_1, W_2, W_3, ..., W_n, the work done in each of these n stages, we have the following expressions.

$$f(m_1) \cdot (x_1 - x_0) \leq W_1 \leq f(M_1) \cdot (x_1 - x_0)$$
$$f(m_2) \cdot (x_2 - x_1) \leq W_2 \leq f(M_2) \cdot (x_2 - x_1)$$
$$\vdots \qquad\qquad \vdots \qquad\qquad \vdots$$
$$f(m_n) \cdot (x_n - x_{n-1}) \leq W_n \leq f(M_n) \cdot (x_n - x_{n-1})$$

Here m_1, m_2, ..., m_n denote values in $[x_0, x_1]$, $[x_1, x_2]$, ..., $[x_{n-1}, x_n]$ at which the force function $f(x) = 2x + 3$ takes on its absolute minimum value in these respective subintervals (in our case $m_1 = x_0$, $m_2 = x_1$, ..., $m_n = x_{n-1}$). M_1, M_2, ..., M_n denote values in $[x_0, x_1]$, $[x_1, x_2]$, ..., $[x_{n-1}, x_n]$ at which $f(x) = 2x + 3$ takes on its absolute maximum value in these respective subintervals (in our case $M_1 = x_1$, $M_2 = x_2$, ..., $M_n = x_n$).

By adding and using the condition

$$W = W_1 + W_2 + \cdots + W_n$$

we obtain

$$\sum_{i=1}^{i=n} f(m_i) \cdot (x_i - x_{i-1}) \leq W \leq \sum_{i=1}^{i=n} f(M_i) \cdot (x_i - x_{i-1})$$

as the bounds for W. Each sequence of partition points $0 = x_0, x_1, x_2, \ldots, x_{n-1}$, $x_n = 6$ yields bounds of this sort for W.

We have seen that by partitioning the interval $[0, 6]$ into more and smaller sub-intervals, we narrow the gap between the lower and upper bounds obtained for W. To close the gap entirely we proceed as we did in the area problem. Since every lower-bound sum, $\Sigma f(m_i) \cdot (x_i - x_{i-1})$, is less than or equal to W, the least upper bound I of the set of all lower-bound sums is less than or equal to W.

$$I = \text{LUB of } \left\{ \sum_{i=1}^{i=n} f(m_i) \cdot (x_i - x_{i-1}) \right\}$$

$$I \leq W$$

Since every upper-bound sum, $\Sigma f(M_i) \cdot (x_i - x_{i-1})$, is greater than or equal to W, the greatest lower bound of the set of all upper-bound sums is greater than or equal to W.

$$J = \text{GLB of } \left\{ \sum_{i=1}^{i=n} f(M_i) \cdot (x_i - x_{i-1}) \right\}$$

$$W \leq J$$

Thus we have

$$I \leq W \leq J.$$

If $I = J$, we define W to be their common value.

More generally, suppose a particle is moved along a line from a position $x = a$ to a position $x = b$ by a variable force whose magnitude at point x in $[a, b]$ is given by the continuous function $y = f(x)$. Suppose the interval $[a, b]$ is partitioned into n subintervals, $[x_0, x_1], [x_1, x_2], \ldots, [x_{n-1}, x_n]$, by choosing a sequence of values $a = x_0, x_1, x_2, \ldots, x_{n-1}, x_n = b$. Also, suppose the total work done W is regarded as being done in n stages, where in the first stage the particle is moved from x_0 to x_1, in the second stage it is moved from x_1 to x_2, and so on. Then we have the following bounds for $W_1, W_2, \ldots, W_n$.

$$f(m_1) \cdot (x_1 - x_0) \leq W_1 \leq f(M_1) \cdot (x_1 - x_0)$$
$$f(m_2) \cdot (x_2 - x_1) \leq W_2 \leq f(M_2) \cdot (x_2 - x_1)$$
$$\vdots \qquad\qquad \vdots \qquad\qquad \vdots$$
$$f(m_n) \cdot (x_n - x_{n-1}) \leq W_n \leq f(M_n) \cdot (x_n - x_{n-1})$$

By adding and using the condition

$$W = W_1 + W_2 + \cdots + W_n$$

we obtain

$$\sum_{i=1}^{i=n} f(m_i) \cdot (x_i - x_{i-1}) \leq W \leq \sum_{i=1}^{i=n} f(M_i) \cdot (x_i - x_{i-1})$$

as the bounds for W, the total work done in moving the particle from a to b. Here $m_1, m_2, \ldots, m_n$ and $M_1, M_2, \ldots, M_n$ denote values in $[x_0, x_1], [x_1, x_2], \ldots, [x_{n-1}, x_n]$ at which $y = f(x)$ takes on its absolute minimum value and absolute maximum value in these respective subintervals. Such bounds for W are obtained for each choice of sequence of values $a = x_0, x_1, x_2, \ldots, x_{n-1}, x_n = b$.

Let I denote the least upper bound of the set of all lower-bound sums.

$$I = \text{LUB of } \left\{ \sum_{i=1}^{i=n} f(m_i) \cdot (x_i - x_{i-1}) \right\}$$

Let J denote the greatest lower bound of the set of all upper-bound sums.

$$J = \text{GLB of } \left\{ \sum_{i=1}^{i=n} f(M_i) \cdot (x_i - x_{i-1}) \right\}$$

If $I = J$, then their common value is defined as the *work done* W by the force in moving the particle from a to b.

Exercises

82. A particle is moved nine feet along a horizontal line by a variable force whose magnitude x feet from the initial position of the particle is given by $f(x) = \sqrt{x} + 1$ pounds.
    a. Determine lower and upper bounds for the work done W by the force on the basis of the sequence of values $\langle 0, 1, 9 \rangle$.
    b. Determine lower and upper bounds for W on the basis of the sequence of values $\langle 0, 1, 4, 9 \rangle$.
83. A particle is moved eight feet along a horizontal line by a variable force whose magnitude x feet from the initial position of the particle is $f(x) = x^2 + 2$ pounds.
    a. Determine lower and upper bounds for the work done W by the force on the basis of the sequence of values $\langle 0, 3, 8 \rangle$.
    b. Determine lower and upper bounds for W on the basis of the sequence of values $\langle 0, 3, 5, 8 \rangle$.
    c. Determine lower and upper bounds for W on the basis of the sequence of values $\langle 0, 2, 3, 5, 8 \rangle$.

28. The definite integral of a function

Although the area and work situations considered in the previous section come out of different backgrounds—geometry and physics—the definitions of area and work that we arrived at are strikingly similar in structure. In both cases we start with a function $y = f(x)$ which is defined and continuous at every value of an interval $[a, b]$. In both situations we introduce a sequence of values $a = x_0, x_1, x_2, \ldots, x_{n-1}, x_n = b$ which partitions the interval $[a, b]$ into subintervals $[x_0, x_1], [x_1, x_2], \ldots,$

$[x_{n-1}, x_n]$. From each such sequence of values we obtain a lower-bound sum,

$$\sum_{i=1}^{i=n} f(m_i) \cdot (x_i - x_{i-1}) = f(m_1) \cdot (x_1 - x_0) + \cdots + f(m_n) \cdot (x_n - x_{n-1})$$

and an upper-bound sum,

$$\sum_{i=1}^{i=n} f(M_i) \cdot (x_i - x_{i-1}) = f(M_1) \cdot (x_1 - x_0) + \cdots + f(M_n) \cdot (x_n - x_{n-1}).$$

Here $m_1, m_2, \ldots, m_n$ are values in $[x_0, x_1], [x_1, x_2], \ldots, [x_{n-1}, x_n]$ at which $y = f(x)$ takes on its absolute minimum value in these respective subintervals. $M_1, M_2, \ldots, M_n$ are values in $[x_0, x_1], [x_1, x_2], \ldots, [x_{n-1}, x_n]$ at which $y = f(x)$ takes on its absolute maximum value in these respective subintervals.

If I is the least upper bound of the set of all lower-bound sums,

$$I = \text{LUB of } \left\{ \sum_{i=1}^{i=n} f(m_i) \cdot (x_i - x_{i-1}) \right\}$$

and J is the greatest lower bound of the set of all upper-bound sums,

$$J = \text{GLB of } \left\{ \sum_{i=1}^{i=n} f(M_i) \cdot (x_i - x_{i-1}) \right\}$$

and

$$I = J$$

then their common value is called the *definite integral of the function $y = f(x)$ over* $[a, b]$ and denoted by

$$\int_a^b f(x) \, dx.$$

We thus see that if $y = f(x)$ is interpreted as a force function, then the definite integral $\int_a^b f(x) \, dx$ expresses the work done by the force in moving a particle from a to b along a line. If $f(x)$ is positive for x between a and b, then $\int_a^b f(x) \, dx$ expresses the area of the region bounded below by the x-axis, on the left and right by the lines $x = a$ and $x = b$, and above by the graph of $y = f(x)$. There are many other situations which exhibit the definite integral structure and we will discuss some of them in the next section.

The definition of $\int_a^b f(x) \, dx$ presupposes that $a < b$. If $a > b$, then $\int_a^b f(x) \, dx$ is defined as $-\int_b^a f(x) \, dx$. Thus $\int_1^0 x^2 \, dx$ is, by definition, $-\int_0^1 x^2 \, dx$. If $a = b$, then $\int_a^b f(x) \, dx = \int_a^a f(x) \, dx$ is defined to be zero. Properties of the definite integral include the following.

1. $\int_a^b K \cdot f(x) \, dx = K \int_a^b f(x) \, dx$, where K is any constant.
2. $\int_a^b [f(x) + g(x)] \, dx = \int_a^b f(x) \, dx + \int_a^b g(x) \, dx.$

The definite integral of a sum of functions is equal to the sum of the definite integrals of the components of the sum, assuming that the definite integrals of the components exist. More generally, this theorem holds for a sum of two or more functions.

3. $\int_a^b f(x)\,dx + \int_b^c f(x)\,dx = \int_a^c f(x)\,dx$ where a, b, and c are any three values.

If b is between a and c (see Figure 23), then the geometric interpretation of this property is that the area under $y = f(x)$ from a to c is the sum of the areas under $y = f(x)$ from a to b and b to c.

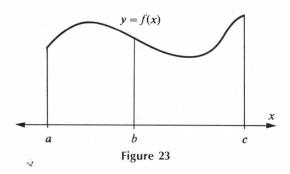

Figure 23

Two basic questions confront us. Under what conditions is $\int_a^b f(x)\,dx$ defined? That is, under what conditions does I, the least upper bound of the set of all lower-bound sums, equal J, the greatest lower bound of the set of all upper-bound sums? Also, when $\int_a^b f(x)\,dx$ is defined, how can its value be determined?

In answer to the first question, continuity of $y = f(x)$ for $a \le x \le b$ suffices to guarantee that $\int_a^b f(x)\,dx$ exists. Insofar as determining the value of $\int_a^b f(x)\,dx$ is concerned, there is a remarkable connection between definite and indefinite integrals which is sometimes called the *fundamental theorem of integral calculus*.

Theorem. If $y = f(x)$ is continuous for $a \le x \le b$ and F is any indefinite integral of f, then

$$\int_a^b f(x)\,dx = F(b) - F(a).$$

That is, to determine $\int_a^b f(x)\,dx$, find any indefinite integral F of f, evaluate F at values a and b, and subtract $F(a)$ from $F(b)$. The difference $F(b) - F(a)$ is frequently denoted by the expression $F(x)]_a^b$.

This connection between definite and indefinite integrals, while most remarkable, is simpler in theory than in practice. Determination of an indefinite integral of a function is rarely a simple matter. Yet when we consider the nature of the definite integral concept, we immediately become grateful for any tools which can help us determine definite integrals. When applicable, the "fundamental theorem" is obviously a powerful ally. Let us first see how it is used and then turn our attention to its proof.

EXAMPLE 10. Determine the integral $\int_2^{10} x^2\,dx$, which expresses the area of the region S bounded by the x-axis, the lines $x = 2$ and $x = 10$, and the graph of $f(x) = x^2$, discussed in the previous section.

Solution. One indefinite integral of $f(x) = x^2$ is $F(x) = \frac{1}{3}x^3$.

$$\int_2^{10} x^2\,dx = \frac{1}{3}x^3\Big]_2^{10} = \frac{1}{3}(10)^3 - \frac{1}{3}(2)^3 = \frac{1000}{3} - \frac{8}{3} = \frac{992}{3}$$

EXAMPLE 11. Determine the integral $\int_0^6 (2x + 3)\,dx$, which expresses the work done W by a variable force defined by $f(x) = 2x + 3$ in moving a particle along a horizontal line from an initial position of $x = 0$ to $x = 6$, discussed in the previous section.

Solution. One indefinite integral of $f(x) = 2x + 3$ is $F(x) = x^2 + 3x$.

$$\int_0^6 (2x + 3)\,dx = (x^2 + 3x)\Big]_0^6 = [(6)^2 + 3(6)] - [(0)^2 + 3(0)] = 54 - 0 = 54$$

The work done W is 54 foot-pounds.

EXAMPLE 12. Determine $\int_1^4 x\sqrt{3 + x^2}\,dx$.

Solution. In Example 1, Section 26 we established that

$$F(x) = \frac{1}{3}(3 + x^2)^{\frac{3}{2}}$$

is an indefinite integral of $f(x) = x\sqrt{3 + x^2}$.

$$\int_1^4 x\sqrt{3 + x^2}\,dx = \frac{1}{3}(3 + x^2)^{\frac{3}{2}}\Big]_1^4$$

$$= \frac{1}{3}(3 + 16)^{\frac{3}{2}} - \frac{1}{3}(3 + 1)^{\frac{3}{2}}$$

$$= \frac{1}{3}(\sqrt{19})^3 - \frac{1}{3}(\sqrt{4})^3$$

$$= \frac{1}{3}[(\sqrt{19})^3 - 8]$$

EXAMPLE 13. Determine $\int_0^5 1000\,e^{-0.1t}\,dt$.

Solution. From Example 2, Section 26 we have that

$$F(t) = -10,000\,e^{-0.1t}$$

is an indefinite integral of $g(t) = 1000\,e^{-0.1t}$.

$$\int_0^5 1000\, e^{-0.1t}\, dt = (-10,000\, e^{-0.1t}) \Big]_0^5$$

$$= -10,000\, e^{-0.5} - (-10,000\, e^0)$$
$$= 10,000(e^0 - e^{-0.5})$$
$$= 10,000(1 - 0.6065)$$
$$= 3935$$

Exercises

Determine the following integrals.

84. $\int_1^2 4x^2\, dx$

85. $\int_0^2 6x^3\, dx$

86. $\int_{-1}^1 (2x^2 + 1)\, dx$

87. $\int_{-2}^3 (3x^2 + 2)\, dx$

88. $\int_0^2 (x^3 + 3)\, dx$

89. $\int_{-2}^4 (4x^2 + 2)\, dx$

90. $\int_{-1}^2 (2x^2 + x - 2)\, dx$

91. $\int_0^3 (3x^2 - 2x + 1)\, dx$

92. $\int_0^{20} (1000 - 2x^2)\, dx$

93. $\int_{-2}^4 (1000 - x^3)\, dx$

94. $\int_0^1 3(3x + 4)^2\, dx$

95. $\int_0^1 2x(x^2 + 1)^4\, dx$

96. $\int_1^2 10x(5x^2 + 2)^3\, dx$

97. $\int_0^2 xe^{-x^2}\, dx$

98. $\int_0^8 500\, e^{-0.5t}\, dt$

99. $\int_0^5 1000\, e^{0.5(5-t)}\, dt$

100. $\int_0^1 xe^x\, dx$

101. $\int_0^{20} (2200 - 100t)\, e^{-0.1t}\, dt$

102. $\int_0^{20} (100 - 20t)\, e^{-0.05t}\, dt$

103. $\int_0^5 \sqrt{4 + x}\, dx$

104. $\int_0^5 (100 + 3t)\, e^{0.1(5-t)}\, dt$

105. $\int_0^9 x\sqrt{x^2 + 16}\, dx$

106. Show that $\int_0^x A\, e^{-rt}\, dt = \dfrac{A}{r}(1 - e^{-rx})$.

107. Show that $\int_0^x A\, e^{r(x-t)}\, dt = \dfrac{A}{r}(e^{rx} - 1)$.

Further discussion of the fundamental theorem

If $y = f(x)$ is continuous for all x satisfying $a \leq x \leq b$ and F is any indefinite integral of f, then

$$\int_a^b f(x)\, dx = F(b) - F(a).$$

To see why this is the case in geometric terms, consider the situation shown in Figure 24, which serves as the basis for our definition of the area function $A(x)$ defined by

$$A(x) = \text{Area of } axst.$$

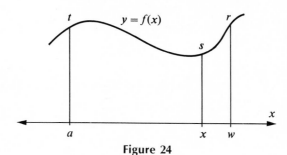

Figure 24

Our first task is to show that $A(x)$ is an indefinite integral of $f(x)$, that is, $A'(x) = f(x)$. Let us begin by observing that

$$A(w) = \text{Area of } awrt.$$

Thus

$$A(w) - A(x) = \text{Area of } xwrs.$$

Let m_w denote a value in $[x,w]$ at which $y = f(x)$ takes on its absolute minimum value, and let M_w denote a value in $[x,w]$ at which $y = f(x)$ takes on its absolute maximum value. Then the area of $xwrs$ is between the areas of the inner and outer rectangles with respective heights $f(m_w)$ and $f(M_w)$ (see Figure 25).

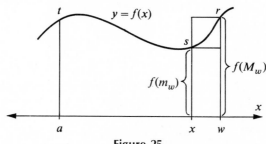

Figure 25

Since the area of *xwrs* is $A(w) - A(x)$ and the areas of the inner and outer rectangles are $f(m_w) \cdot (w - x)$ and $f(M_w) \cdot (w - x)$, respectively, we have

$$f(m_w) \cdot (w - x) \le A(w) - A(x) \le f(M_w) \cdot (w - x).$$

By dividing by $w - x$, which is positive, we obtain

$$f(m_w) \le \frac{A(w) - A(x)}{w - x} \le f(M_w).$$

Let us now determine what happens as $w \to x$. As $w \to x$, m_w, which is trapped between x and w, approaches x, and $f(m_w) \to f(x)$ (since f is continuous at x). For the same reasons, $f(M_w) \to f(x)$.

Since $[A(w) - A(x)]/(w - x)$ is trapped between two quantities, $f(m_w)$ and $f(M_w)$, which approach $f(x)$, it too must approach $f(x)$. Like the passenger in a crowded New York City subway car who is trapped between two fellows heading for the exit, it has no choice. It must go where they go. Thus we have

$$\underset{w \to x}{\text{limit}} \, \frac{A(w) - A(x)}{w - x} = f(x).$$

But, by definition of derivative,

$$\underset{w \to x}{\text{limit}} \, \frac{A(w) - A(x)}{w - x} = A'(x).$$

Thus we have that

$$A'(x) = f(x)$$

which means that $A(x)$ is an indefinite integral of $f(x)$. This concludes the first part of our geometrically-flavored demonstration of the fundamental theorem.

The second part is short and simple. Since area is a definite integral concept, for any value $b > a$ we have

$$A(b) = \int_a^b f(x) \, dx.$$

Also,

$$A(a) = \int_a^a f(x) \, dx = 0.$$

Thus

$$A(b) - A(a) = \int_a^b f(x) \, dx.$$

Let F denote any other indefinite integral of f. That is, $F' = f$. Then

$$F(x) = A(x) + C$$

for a suitable constant C. It thus follows that

$$F(b) - F(a) = (A(b) + C) - (A(a) + C)$$
$$= A(b) - A(a)$$
$$= \int_a^b f(x)\, dx$$

and our demonstration is completed.

29. Concepts with an integral structure

Now that we have developed a basic tool to help us determine definite integrals, let us consider some of the concepts which have an integral structure and apply the results obtained to their study.

Area of a region

The area of the region S bounded by the graph of $y = f(x)$, the x-axis, and the lines $x = a$ and $x = b$ is defined as

$$A(S) = \int_a^b f(x)\, dx$$

provided that $f(x)$ is positive for x between a and b. If $f(x)$ is negative for x between a and b, then $A(S)$ is defined as

$$A(S) = -\int_a^b f(x)\, dx.$$

The area of the region S between the graphs of $y = f(x)$ and $y = g(x)$, and the lines $x = a$ and $x = b$ (see Figure 26) is defined as the difference in the areas under the two curves.

$$A(S) = \int_a^b f(x)\, dx - \int_a^b g(x)\, dx = \int_a^b [f(x) - g(x)]\, dx$$

Here it is assumed that, for x between a and b, $f(x)$ and $g(x)$ are both positive and $g(x) < f(x)$.

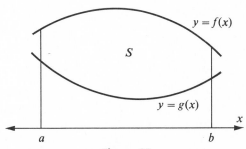

Figure 26

EXAMPLE 14. Find the area $A(S)$ of the region S bounded by the x-axis, the lines $x = -2$ and $x = 3$, and the graph of $f(x) = 3x^2 + 2$ (see Figure 27).

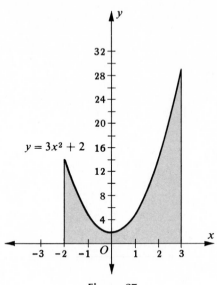

Figure 27

Solution.

$$A(S) = \int_{-2}^{3} (3x^2 + 2)\, dx = (x^3 + 2x) \Big]_{-2}^{3}$$

$$= 33 - (-12) = 45$$

EXAMPLE 15. Find the area $A(S)$ of the region S bounded by the graphs of $f(x) = x + 6$ and $g(x) = x^2 + 1$, and the lines $x = -1$ and $x = 2$ (see Figure 28 at the top of the next page).

Solution.

$$A(S) = \int_{-1}^{2} [f(x) - g(x)]\, dx = \int_{-1}^{2} (x + 6 - x^2 - 1)\, dx$$

$$= \int_{-1}^{2} (x + 5 - x^2)\, dx$$

$$= \left(\frac{1}{2} x^2 + 5x - \frac{1}{3} x^3 \right) \Big]_{-1}^{2}$$

$$= \frac{28}{3} - \left(-\frac{25}{6} \right) = \frac{71}{6}$$

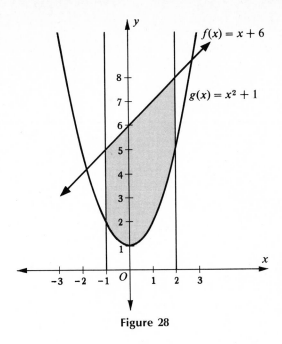

Figure 28

Exercises

For each exercise, draw a diagram, shade the region S, and determine $A(S)$.

108. S is bounded by the x-axis, the lines $x = 0$ and $x = 4$, and the graph of $f(x) = -x^2 + 16$.

109. S is bounded by the x-axis, the lines $x = -1$ and $x = 3$, and the graph of $f(x) = 2x^2 + 4$.

110. S is bounded by the x-axis, the lines $x = -2$ and $x = 4$, and the graph of $f(x) = 2x^2 + x + 3$.

111. S is bounded by the x-axis, the lines $x = -3$ and $x = 0$, and the graph of $f(x) = x^3$.

112. S is bounded by the x-axis, the lines $x = 1$ and $x = 4$, and the graph of $f(x) = x^3 + 2$.

113. S is bounded by the graphs of $f(x) = -x^2 + 16$ and $g(x) = 2x^2 + 4$, and the lines $x = -1$ and $x = 2$.

114. S is bounded by the graphs of $f(x) = x + 4$ and $g(x) = \frac{1}{2}x^2$, and the lines $x = -1$ and $x = 4$.

115. S is bounded by the graphs of $f(x) = -x^2 + 16$ and $g(x) = x^3$, and the lines $x = 0$ and $x = 2$.

116. S is bounded by the graphs of $f(x) = \sqrt{x}$ and $g(x) = 1/x$, and the line $x = 4$.

117. S is bounded by the graphs of $f(x) = \sqrt{x}$ and $g(x) = x$.

118. S is bounded by the graphs of $f(x) = \sqrt{x}$ and $g(x) = x^3$.

119. S is bounded by the graphs of $f(x) = \ln x$, the x-axis, and the line $x = e$.

120. S is bounded by the graphs of $f(x) = e^x$ and $g(x) = e^{-x}$, and the line $x = 3$.

Consumers' surplus

Consider the demand function $x = g(p)$ which expresses x, the amount of a commodity bought by the market per unit time, as a function of p, the unit price of the commodity. The inverse function $p = f(x)$ expresses price p as a function of quantity x. If the market price is p_1 and the corresponding demand is $x_1 = g(p_1)$, then those consumers who would be willing to pay more for the commodity than market price p_1 gain from the market price being set at p_1.

The area of the region bounded below by the line $p = p_1$, on the left and right by the p-axis ($x = 0$) and the line $x = x_1$, and above by the graph of the price-quantity demand function $p = f(x)$ (see Figure 29), serves, under certain economic assumptions, as a numerical measure of the consumers' total gain. The area of this region, called the *consumers' surplus* for price p_1, is defined by

$$\left(\int_0^{x_1} f(x)\, dx \right) - p_1 x_1.$$

The product $p_1 x_1$, of price p_1 and quantity x_1, bought at this price expresses the amount paid by consumers for the commodity. This amount is represented geometrically by the area of the rectangle with length x_1 and height p_1 (see Figure 29).

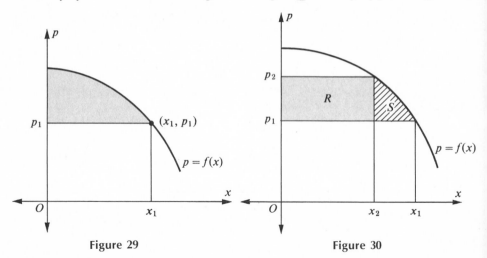

| Figure 29 | Figure 30 |

If the price changes from p_1 to p_2, with a corresponding change in the quantity demanded from $x_1 = g(p_1)$ to $x_2 = g(x_2)$, then the change in consumers' surplus ΔC is the sum of the areas of regions R and S shown in Figure 30. If the price rises from p_1 to p_2, ΔC serves as a numerical measure of the injury suffered by consumers. If the price falls from p_2 to p_1, ΔC is the benefit derived by consumers.

The areas of R and S are

$$A(R) = p_2 x_2 - p_1 x_2$$

$$A(S) = \int_{x_2}^{x_1} f(x)\, dx - (p_1 x_1 - p_1 x_2) = \int_{x_2}^{x_1} f(x)\, dx - p_1 x_1 + p_1 x_2.$$

Thus the sum $A(R) + A(S)$ is

$$A(R) + A(S) = p_2 x_2 - p_1 x_2 + \int_{x_2}^{x_1} f(x)\, dx - p_1 x_1 + p_1 x_2$$

$$= \int_{x_2}^{x_1} f(x)\, dx - p_1 x_1 + p_2 x_2$$

$$= \int_{x_2}^{x_1} f(x)\, dx - (p_1 x_1 - p_2 x_2).$$

We thereby obtain the following results for $\Delta C = A(R) + A(S)$.*

$$\Delta C = \int_{x_2}^{x_1} f(x)\, dx - (p_1 x_1 - p_2 x_2)$$

$$\Delta C = \int_{x_2}^{x_1} f(x)\, dx - \text{(change in expenditure)}$$

EXAMPLE 16. The demand functions for tea in a certain market are $p = 1000 - 2x^2$ and $x = \sqrt{(1000 - p)/2}$, where p is the price in dollars per ton and x is the number of tons bought per month. Determine the consumers' surplus for the price of $200 per ton. If the price changes from $118 to $200 per ton, find the change in the consumers' surplus.

Solution. For $p = 200$, $x = \sqrt{(1000 - 200)/2} = 20$. Thus the consumers' surplus for a price of $200 a ton is

$$\int_0^{20} (1000 - 2x^2)\, dx - 200(20) = \left(1000x - \frac{2}{3} x^3 \right)\Big]_0^{20} - 4000$$

$$= \frac{44{,}000}{3} - 4000$$

$$= 10{,}666\,\frac{2}{3}.$$

For $p = 118$, $x = 21$, and for $p = 200$, $x = 20$. Thus the change in consumers' surplus ΔC for a price change from $118 to $200 a ton is

$$\Delta C = \int_{20}^{21} (1000 - 2x^2)\, dx - [118(21) - 200(20)]$$

$$= \left(1000x - \frac{2}{3} x^3 \right)\Big]_{20}^{21} + 1522$$

$$= \left(\frac{44{,}478}{3} - \frac{32{,}000}{3} \right) + 1522$$

$$= 5681\,\frac{1}{3}.$$

*For further discussion of the concept of consumers' surplus, see, for example, D. S. Watson, *Price Theory and Its Uses,* 2nd ed. (Boston: Houghton Mifflin, 1968), pp. 68–71, 139–140, 317, 326–328.

If the price increases from \$118 to \$200 per ton, then $\Delta C = 5681\frac{1}{3}$ serves as a numerical measure of the injury suffered by consumers. If the price drops from \$200 to \$118, $\Delta C = 5681\frac{1}{3}$ serves as a numerical measure of the benefit derived by consumers.

Exercises

121. The demand functions for a certain commodity are $p = 500 - x^2$ and $x = \sqrt{500 - p}$.
    a. Determine the consumers' surplus for $p = 100$.
    b. If p drops from 100 to 59, what is the change in consumers' surplus?
    c. If p rises from 100 to 139, what is the change in consumers' surplus?
122. The demand functions for a certain commodity are $p = 2000 - 2x^2$ and $x = \sqrt{(2000 - p)/2}$.
    a. Determine the consumers' surplus for $p = 200$.
    b. If p drops from 200 to 78, what is the change in consumers' surplus?
    c. If p rises from 200 to 318, what is the change in consumers' surplus?
123. The total cost function of a coffee-producing monopolist is $c(x) = \frac{19}{2}x^2 + 10x + 500$ dollars for an output of x tons a week. The demand functions for coffee in the market supplied are $p = (50 - x)^2$ and its inverse $x = 50 - \sqrt{p}$, where p is the price of coffee in dollars per ton and x is the output in tons per week. Find the consumers' surplus for the market price determined by the given conditions. Hint: Determine the output for which the monopolist's profit is maximized (see Section 21).

Producers' surplus

Consider the supply functions $x = s(p)$, which expresses x, the amount of a commodity that producers wish to sell to a market per unit time, as a function of its market price p, and the inverse function $p = h(x)$, which expresses market price as a function of quantity supplied. If the market price of the commodity is p_1 and the corresponding amount supplied to the market is $x_1 = s(p_1)$, then those producers who would be willing to supply the commodity to the market when the price is less than p_1 gain from the market price being set at p_1.

The area of the region bounded below by the price-quantity supply function $p = h(x)$, above by the line $p = p_1$, and on the left and right by the p-axis ($x = 0$) and the line $x = x_1$ (see Figure 31), serves, under certain economic assumptions, as a numerical measure of the total producer gain. The area of this region, called the *producers' surplus*, is given by

$$p_1 x_1 - \int_0^{x_1} h(x)\, dx.$$

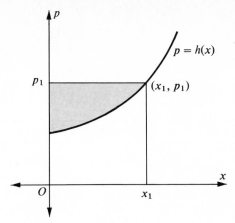

Figure 31

EXAMPLE 17. The price-quantity supply function for sugar in a certain market is $p = x^2$, where x is tons of sugar supplied per month and p is the market price of sugar in dollars per ton. Determine the producers' surplus corresponding to a market price of $100 per ton.

Solution. For $p = 100$, $x = 10$. Thus the producers' surplus is

$$100(10) - \int_0^{10} x^2 \, dx = 1000 - \left(\frac{1}{3} x^3 \right) \bigg]_0^{10}$$

$$= 1000 - \frac{1000}{3}$$

$$= \frac{2000}{3}.$$

Exercises

124. The price-quantity supply function for a certain commodity is $p = 2x^2$.
   a. Find the producers' surplus for $p = 200$.
   b. Find the producers' surplus for $p = 162$.
125. The price-quantity supply function for a certain commodity is $p = \frac{1}{4}x^3$.
   a. Find the producers' surplus for $p = 16$.
   b. Find the producers' surplus for $p = 54$.

Present value of an income stream

Suppose an investment made now yields incomes of $1000 one year later, $2000 two years later, and $2000 three years later. If the going interest rate is 6% per annum, then in the mathematics of finance (see Appendix 2) it is shown that the respective present values of these incomes are

$$1000(1.06)^{-1} = \$943.40$$
$$2000(1.06)^{-2} = \$1780.00$$
$$2000(1.06)^{-3} = \$1679.24.$$

Thus the total present worth of this discrete income stream is the sum $943.40 + $1780.00 + $1679.24 = $4402.64. In general, the total present value of a discrete income stream consisting of the amounts $A_1, A_2, \ldots, A_n$ is found by determining the present values of $A_1, A_2, \ldots, A_n$ and adding.

But suppose income is obtained continuously over time at a rate of $f(t)$ dollars per year t years from the present ($t = 0$). Then the present value of the income stream is a definite integral concept defined in the following way. If income is obtained continuously from time $t = 0$ to time $t = x$ years at the rate of r per annum compounded continuously, then the *present value* Π of the income stream is defined by

$$\Pi = \int_0^x f(t)e^{-rt}\, dt.$$

EXAMPLE 18. If income from a certain investment is derived at a constant rate of $1000 per year for five years and interest is 10% per annum compounded continuously, then the present value Π of the income stream is

$$\Pi = \int_0^5 1000\, e^{-0.1t}\, dt.$$

From Example 2, Section 26, we have that $F(t) = -10,000\, e^{-0.1t}$ is an indefinite integral of $g(t) = 1000\, e^{-0.1t}$.

$$\Pi = (-10,000\, e^{-0.1t})\Big]_0^5 = -10,000\, e^{-0.5} - (-10,000\, e^0)$$
$$= 10,000(e^0 - e^{-0.5})$$
$$= 10,000(1 - 0.6065)$$
$$= \$3935$$

A problem in equipment-investment analysis

An important problem in equipment-investment analysis is concerned with determining the present value to a firm of a piece of production equipment which earns revenue at the continuous rate of $R(t)$ dollars per year t years after its installation and incurs operation and maintenance expenses at the continuous rate of $E(t)$ dollars per year t years after its installation (this does not include depreciation and interest on investment). Income is derived from such a piece of equipment at the continuous rate of $f(t) = R(t) - E(t)$ dollars per year t years after its installation.

If cost after installation and salvage value considerations are ignored for the moment, then the present value of a piece of equipment which is assumed to have a useful life of x years is

$$\Pi = \int_0^x f(t)e^{-rt}\, dt$$

where the interest rate is r per annum compounded continuously.

To illustrate, consider a firm that has acquired a machine which is assumed to have a useful life of twenty-years. It is estimated that net income is derived at the continuous rate of

$$f(t) = 2200 - 100t$$

dollars t years after installation. If the interest on investment is 10% per annum compounded continuously, then, ignoring installation costs and salvage value considerations for the moment, the present value of the equipment of the firm is

$$\Pi = \int_0^{20} (2200 - 100t)e^{-0.1t}\, dt.$$

Now

$$F(t) = -12{,}000\, e^{-0.1t} + 1000te^{-0.1t}$$

is an indefinite integral of $g(t) = (2200 - 100t)e^{-0.1t}$ (see Example 6, Section 26).

$$\Pi = (-12{,}000\, e^{-0.1t} + 1000te^{-0.1t})\Big]_0^{20}$$
$$= (-12{,}000\, e^{-2} + 20{,}000\, e^{-2}) - (-12{,}000\, e^0)$$
$$= 8000\, e^{-2} + 12{,}000\, e^0$$

Since $e^{-2} = 0.13534$ and $e^0 = 1$, we obtain

$$\Pi = \$13{,}082.72.$$

As we have observed, the expression

$$\Pi = \int_0^x f(t)e^{-rt}\, dt$$

for the present value to a firm of a piece of production equipment does not take the cost after installation and salvage value of the equipment into consideration. The cost C of the equipment after installation is simple to deal with. Just subtract it from Π. If S is the salvage value of the equipment x years later, then

$$S e^{-rx}$$

is the present value of S, where the interest rate is r per annum compounded continuously (see Section 7).

Thus the present value Π to a firm of a piece of production equipment which earns net revenue (total revenue minus the maintenance expenses incurred) at the continuous rate of $f(t)$ dollars per year t years after its installation is given by

$$\Pi = \int_0^x f(t)e^{-rt}\,dt + Se^{-rx} - C$$

where x is the life of the equipment in years, S is its salvage value after x years, C is the cost of the equipment after installation, and r is the continuously compounded annual interest rate.*

In our illustrative example, if the installed cost of the equipment is $1000 and its salvage value after twenty years is $200, then, after installed cost and salvage value have been taken into account, the present value of the equipment is

$$\begin{aligned}
\Pi &= \$13{,}082.72 + 200\,e^{-2} - \$1000 \\
&= \$13{,}082.72 + 200(0.13534) - \$1000 \\
&= \$12{,}109.79.
\end{aligned}$$

Exercises

126. If income from a certain investment is obtained continuously at the constant rate of $2000 per year for ten years, and interest is 10% per annum compounded continuously, find the present value of the income stream ($e^{-1} = 0.3679$).
127. If income from a certain investment is obtained continuously at the constant rate of $5000 per year for six years, and interest is 12% per annum compounded continuously, find the present value of the income stream ($e^{-0.72} = 0.4916$).
128. If income from a certain investment is obtained continuously at the constant rate of $3000 per year for eight years, and interest is 8% per annum compounded continuously, find the present value of the income stream ($e^{-0.64} = 0.5273$).
129. Show that if income from an investment is obtained continuously at the constant rate of $\$A$ per year for x years, and interest is r per annum compounded continuously, then the present value of the income stream is $\dfrac{A}{r}\,(1 - e^{-rx})$ dollars.
130. A machine acquired by a firm is assumed to have a useful life of 10 years. It is estimated that net income is derived at the continuous rate of $f(t) = 1800 - 50t$ dollars per year t years after installation. Interest on investment is 10% per annum compounded continuously.
    a. If installation costs and salvage value considerations are ignored, determine the present value of the machine ($e^{-1} = 0.36788$).
    b. If the installed cost of the machine is $800 and its salvage value at the end of 10 years is $100, determine the present value of the machine.
131. A piece of production equipment acquired by a company is assumed to have

*For further discussion of equipment-investment analysis, see such works as E. H. Bowman and R. B. Fetter, *Analysis for Production and Operations Management,* 3rd ed. (Homewood, Illinois: Richard P. Irwin, Inc., 1967), Chapter 10; F. and V. Lutz, *The Theory of Investment of the Firm* (Princeton, New Jersey: Princeton University Press, 1951), Chapter VIII; and G. A. D. Preinreich, "The Economic Life of Industrial Equipment," *Econometrica* (January 1940), pp. 12–44.

a useful life of 20 years. It is estimated that net income is derived at the continuous rate of $f(t) = 1000 - 20t$ dollars per year t years after installation. Interest on investment is 5% per annum compounded continuously.

a. If installation costs and salvage value considerations are ignored, determine the present value of the production equipment.
b. If the installed cost of the production equipment is $500 and its salvage value at the end of 20 years is $50, determine the present value of the equipment.

Future value of an investment stream

Suppose $1000 is invested now and $2000, $3000, and $4000 are invested in successive years at the rate of 6% compounded annually. In the mathematics of finance (see Appendix 2) it is shown that at the end of the investment period (three years from now) these respective amounts will be worth the following.

$$1000(1.06)^3 = 1000(1.1910) = \$1191.00$$
$$2000(1.06)^2 = 2000(1.1236) = \$2247.20$$
$$3000(1.06)^1 = 3000(1.0600) = \$3180.00$$
$$4000(1.06)^0 = 4000(1) \qquad = \$4000.00$$

Thus the total value of this discrete investment stream at the end of the investment period is the sum $\$1191.00 + \$2247.20 + \$3180.00 + \$4000.00 = \$10,618.20$.

More generally, if $\$A_0, \$A_1, \$A_2, \ldots, \$A_n$ are invested now and in succeeding years at the rate of r compounded annually, the total value of this discrete investment stream at the end of the investment period (n years from now) is given by the sum

$$\sum_{i=0}^{i=n} A_i(1 + r)^{n-i} = A_0(1 + r)^n + A_1(1 + r)^{n-1} + \cdots + A_n.$$

If investment is made continuously over time at a rate of $f(t)$ dollars per year t years from the present ($t = 0$) at a rate of r per annum compounded continuously, then the total sum accumulated x years from now is defined as

$$S = \int_0^x f(t)e^{r(x-t)} \, dt.$$

EXAMPLE 19. If money is invested continuously at the constant rate of $1000 per year at an interest rate of 10% per annum compounded continuously, then the total sum accumulated at the end of five years is

$$S = \int_0^5 1000 \, e^{0.1(5-t)} \, dt.$$

From Example 3, Section 26 we have that

$$F(t) = -10,000 \, e^{0.1(5-t)}$$

is an indefinite integral of $g(t) = 1000 \, e^{0.1(5-t)}$.

$$S = (-10,000 \, e^{0.1(5-t)}) \Big]_0^5$$

$$= -10,000 \, e^0 - 10,000 \, e^{0.5}$$

$$= 10,000 \, (e^{0.5} - e^0)$$
$$= 10,000(1.6487 - 1)$$
$$= \$6487.00$$

Exercises

132. If money is invested continuously at the constant rate of \$2000 per year at an interest rate of 10% per annum compounded continuously, find the total sum accumulated at the end of ten years.

133. If money is invested continuously at the constant rate of \$3000 per year at an interest rate of 12% per annum compounded continuously, find the total sum accumulated at the end of eight years ($e^{0.96} = 2.6117$).

134. If money is invested continuously at the constant rate of \$5000 per year at an interest rate of 8% per annum compounded continuously, find the total sum accumulated at the end of eleven years ($e^{0.88} = 2.4109$).

135. Show that if money is invested continuously at the constant rate of \$A per year at an interest rate of r per annum compounded continuously, then the total sum accumulated at the end of x years is $\frac{A}{r}(e^{rx} - 1)$ dollars.

136. If money is invested continuously at the rate $f(t) = 100 + 3t$ dollars per year at an interest rate of 10% per annum compounded continuously, find the total sum accumulated at the end of five years ($e^{0.5} = 1.6487$).

137. If money is invested continuously at the rate $f(t) = 500 - 5t$ dollars per year at an interest rate of 12% per year compounded continuously, find the total sum accumulated at the end of eight years.

Work done by a variable force

Suppose a particle is moved along a line from $x = a$ to $x = b$ by a variable force whose magnitude at value x in $[a, b]$ is given by the continuous function $y = f(x)$. Then the *work done* by the force is defined as

$$W = \int_a^b f(x) \, dx.$$

EXAMPLE 20. A particle is moved six feet from $x = 2$ to $x = 8$ along a line by a force whose magnitude at value x in $[2, 8]$ is $f(x) = 4x + 3$ pounds. The work done by this force is the following.

$$W = \int_2^8 (4x + 3)\, dx = (2x^2 + 3x)\Big]_2^8 = 152 - 14 = 138 \text{ foot-pounds}$$

Exercises

138. A particle is moved ten feet from $x = 1$ to $x = 11$ along a line by a force whose magnitude at value x in $[1, 11]$ is $f(x) = x^2 + 2x + 1$ pounds. Determine the work done by this force.
139. A particle is moved nine feet from $x = 0$ to $x = 9$ along a line by a force whose magnitude at value x in $[0, 9]$ is $f(x) = \sqrt{x} + 1$ pounds. Determine the work done by this force.
140. A particle is moved four feet from $x = 3$ to $x = 7$ along a line by a force whose magnitude at value x in $[3, 7]$ is $f(x) = 5x^4 + 8$ pounds. Determine the work done by this force.

Integral representations for old faces

By means of the fundamental theorem of integral calculus, such notions as total cost and total revenue of a firm can be expressed in terms of definite integrals of their corresponding marginal concepts. If $f(x)$ is the marginal cost function of a firm, then the total cost function $c(x)$ is an indefinite integral of $f(x)$ since, by definition, $f(x) = c'(x)$. From the fundamental theorem of integral calculus we have

$$c(x) - c(0) = \int_0^x f(x)\, dx.$$

Thus

$$c(x) = c(0) + \int_0^x f(x)\, dx. \tag{1}$$

$c(0)$ represents the firm's cost when there is no output, that is, the fixed cost of the firm irrespective of output.

Equation (1) gives us a view of total cost in terms of fixed and marginal costs. It simply says that the total cost for output x is fixed cost plus the integral of the marginal cost function from 0 to x. Geometrically this says that the total cost for output x is the fixed cost plus the area under the marginal cost curve from 0 to x (see Figure 32 on the next page).

To illustrate, consider an aluminum producer whose fixed cost is \$1000 per week and whose marginal cost for an output of x tons per week is

$$f(x) = x^2 - 18x + 100.$$

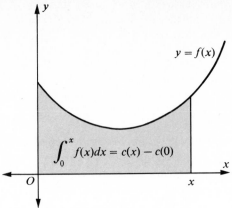

Figure 32

Then his total cost for output x is

$$c(x) = 1000 + \int_0^x (x^2 - 18x + 100)\, dx$$

that is, $c(x)$ equals 1000 plus the area under the graph of $f(x) = x^2 - 18x + 100$ between 0 and x. Thus, for example, $c(10)$, the total cost of producing ten tons of aluminum per week, is

$$c(10) = 1000 + \int_0^{10} (x^2 - 18x + 100)\, dx$$

$$= 1000 + \left(\frac{1}{3}x^3 - 9x^2 + 100x\right)\Big]_0^{10}$$

$$= 1000 + \frac{1300}{3}$$

$$= \$1433.33.$$

If $f(x)$ is the marginal revenue function of a firm, then by an analogous argument we obtain

$$R(x) = R(0) + \int_0^x f(x)\, dx$$

for the total revenue function $R(x)$. Since $R(0) = 0$ (no production, no revenue), we have

$$R(x) = \int_0^x f(x)\, dx$$

that is, the total revenue derived from the production and sale of x units equals the area under the marginal revenue curve between 0 and x (see Figure 33).

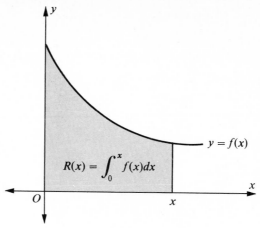

Figure 33

30. A limit formulation of the definite integral concept

In certain situations it is convenient to express the definite integral of a function as a limit rather than as a least upper bound or greatest lower bound. In this section we will discuss how this is done.

To define the definite integral of $y = f(x)$ over interval $[a, b]$, $[a, b]$ is partitioned into n subintervals $[a, x_1], [x_1, x_2], \ldots, [x_{n-1}, b]$ by choosing a sequence of values $a = x_0, x_1, x_2, \ldots, x_{n-1}, x_n = b$ (see Figure 34). From each such sequence we obtain a lower-bound sum,

$$\sum_{i=1}^{i=n} f(m_i) \cdot (x_i - x_{i-1}) = f(m_1) \cdot (x_1 - x_0) + \cdots + f(m_n) \cdot (x_n - x_{n-1})$$

and an upper-bound sum,

$$\sum_{i=1}^{i=n} f(M_i) \cdot (x_i - x_{i-1}) = f(M_1) \cdot (x_1 - x_0) + \cdots + f(M_n) \cdot (x_n - x_{n-1}).$$

Here $m_1, m_2, \ldots, m_n$ are values in $[x_0, x_1], [x_1, x_2], \ldots, [x_{n-1}, x_n]$ at which $y = f(x)$ takes on its absolute minimum value in these respective subintervals. $M_1, M_2, \ldots, M_n$ are values in $[x_0, x_1], [x_1, x_2], \ldots, [x_{n-1}, x_n]$ at which $y = f(x)$ takes on its absolute maximum value in these respective subintervals.

To bridge the gap between the lower-bound sums and upper-bound sums in least-upper-bound and greatest-lower-bound terms, we introduced I, the least upper bound

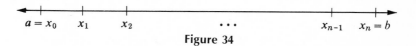

Figure 34

of the set of all lower-bound sums,

$$I = \text{LUB of} \left\{ \sum_{i=1}^{i=n} f(m_i) \cdot (x_i - x_{i-1}) \right\}$$

and J, the greatest lower bound of the set of all upper-bound sums,

$$J = \text{GLB of} \left\{ \sum_{i=1}^{i=n} f(M_i) \cdot (x_i - x_{i-1}) \right\}.$$

If $I = J$, their common value is the definite integral of $y = f(x)$ over $[a,b]$.

Another way of bridging the gap between the lower-bound sums and upper-bound sums is suggested by the observation that the gap gets smaller as you choose sequences $a = x_0, x_1, x_2, \ldots, x_{n-1}, x_n = b$ with more and more values in such a way that the distance between successive values (x_0 and x_1, x_1 and x_2, and so on) gets smaller and smaller.

For example, recall the problem of defining the area $A(S)$ of the region S bounded by the x-axis, the graph of $f(x) = x^2$, and the lines $x = 2$ and $x = 10$ (see Section 27). We saw that the sequence $<2, 5, 10>$ yields the lower sum 137 and upper sum 575 (with a gap of 438); the sequence $<2, 4, 5, 10>$ yields the lower sum 149 and upper sum 557 (with a gap of 408); and the sequence $<2, 4, 5, 8, 10>$ yields the lower sum 227 and upper sum 449 (with a gap of 222). While there is still a long way to go to close the gap completely, the idea suggested is that the behavior of the lower-bound and upper-bound sums be examined as we choose more and more values for our partitioning sequences $a = x_0, x_1, x_2, \ldots, x_{n-1}, x_n = b$. And we will want to choose these values in such a way that the distance between consecutive values ($\Delta x_1 = x_1 - x_0$, $\Delta x_2 = x_2 - x_1$, and so on) approaches zero.

If under these conditions ($n \to \infty$, $\Delta x_i = x_i - x_{i-1}$, and $\Delta x_i \to 0$), the corresponding lower-bound sums

$$\sum_{i=1}^{i=n} f(m_i) \cdot (x_i - x_{i-1})$$

approach a single number I, then we say that I is the limit of the lower sums and write

$$\lim_{\substack{n \to \infty \\ \Delta x_i \to 0}} \sum_{i=1}^{i=n} f(m_i) \cdot (x_i - x_{i-1}) = I.$$

If, under the same conditions, the corresponding upper-bound sums

$$\sum_{i=1}^{i=n} f(M_i) \cdot (x_i - x_{i-1})$$

approach a single number J, then we say that J is the limit of the upper-bound sums and write

$$\underset{\substack{n \to \infty \\ \Delta x_i \to 0}}{\text{limit}} \sum_{i=1}^{i=n} f(M_i) \cdot (x_i - x_{i-1}) = J.$$

If $I = J$, then their common value is defined as the *definite integral of $y = f(x)$ over* $[a, b]$.

This definition is equivalent to the previous one given. I and J, defined as a limit, are equal to the least upper bound of the set of all lower-bound sums and greatest lower bound of the set of all upper-bound sums, respectively.

In the formulations of the definite integral concept that have been considered, lower-bound and upper-bound sums

$$\sum_{i=1}^{i=n} f(m_i) \cdot (x_i - x_{i-1}) \qquad \text{and} \qquad \sum_{i=1}^{i=n} f(M_i) \cdot (x_i - x_{i-1})$$

were formed by choosing values $m_1, m_2, \ldots, m_n$ and $M_1, M_2, \ldots, M_n$ in the subintervals $[x_0, x_1], [x_1, x_2], \ldots, [x_{n-1}, x_n]$ at which $y = f(x)$ takes on absolute minimum and absolute maximum values in these respective subintervals.

Another approach is to arbitrarily choose values $c_1, c_2, \ldots, c_n$ in $[x_0, x_1], [x_1, x_2]$, $\ldots, [x_{n-1}, x_n]$, respectively, form the more general sum

$$\sum_{i=1}^{i=n} f(c_i) \cdot (x_i - x_{i-1})$$

and examine its limit behavior as $n \to \infty$ and $\Delta x_i \to 0$, $\Delta x_i = x_i - x_{i-1}$. That is, examine its limit behavior as we choose more and more values for our partitioning sequences $a = x_0, x_1, x_2, \ldots, x_{n-1}, x_n = b$ and choose these values in such a way that the distance between consecutive values approaches zero.

If as partitioning sequences are chosen in the above manner, the sums

$$\sum_{i=1}^{i=n} f(c_i) \cdot (x_i - x_{i-1})$$

approach a single number K, then we say that K is the limit of these sums and write

$$\underset{\substack{n \to \infty \\ \Delta x_i \to 0}}{\text{limit}} \sum_{i=1}^{i=n} f(c_i) \cdot (x_i - x_{i-1}) = K.$$

If this limit exists, then K is defined as the *definite integral of $y = f(x)$ over* $[a, b]$.

It is instructive and not difficult to see that $K = I = J$ and that this formulation of the definite integral concept is equivalent to the limit formulation just given.

Proof. Part 1. Suppose

$$\underset{\substack{n \to \infty \\ \Delta x_i \to 0}}{\text{limit}} \sum_{i=1}^{i=n} f(c_i) \cdot (x_i - x_{i-1}) = K$$

where $c_1, c_2, \ldots, c_n$ are any values in $[x_0, x_1], [x_1, x_2], \ldots, [x_{n-1}, x_n]$, respectively. If $c_1 = m_1, c_2 = m_2, \ldots, c_n = m_n$ are chosen, then we obtain

$$\underset{\substack{n \to \infty \\ \Delta x_i \to 0}}{\text{limit}} \sum_{i=1}^{i=n} f(m_i) \cdot (x_i - x_{i-1}) = K.$$

Thus $I = K$.

If $c_1 = M_1, c_2 = M_2, \ldots, c_n = M_n$ are chosen, then we obtain

$$\underset{\substack{n \to \infty \\ \Delta x_i \to 0}}{\text{limit}} \sum_{i=1}^{i=n} f(M_i) \cdot (x_i - x_{i-1}) = K.$$

Thus $J = K$.

Part 2. Suppose

$$\underset{\substack{n \to \infty \\ \Delta x_i \to 0}}{\text{limit}} \sum_{i=1}^{i=n} f(m_i) \cdot (x_i - x_{i-1}) = K$$

and

$$\underset{\substack{n \to \infty \\ \Delta x_i \to 0}}{\text{limit}} \sum_{i=1}^{i=n} f(M_i) \cdot (x_i - x_{i-1}) = K.$$

We want to show that under these conditions

$$\underset{\substack{n \to \infty \\ \Delta x_i \to 0}}{\text{limit}} \sum_{i=1}^{i=n} f(c_i) \cdot (x_i - x_{i-1}) = K.$$

Since $m_1, m_2, \ldots, m_n$ and $M_1, M_2, \ldots, M_n$ are values at which absolute minimum and absolute maximum values are taken on by $y = f(x)$ in $[x_0, x_1], [x_1, x_2], \ldots, [x_{n-1}, x_n]$, respectively, we have the following results.

$$f(m_1) \leq f(c_1) \leq f(M_1)$$
$$f(m_2) \leq f(c_2) \leq f(M_2)$$
$$\vdots \qquad \vdots \qquad \vdots$$
$$f(m_n) \leq f(c_n) \leq f(M_n)$$

Multiplying by $x_1 - x_0, x_2 - x_1, \ldots, x_n - x_{n-1}$, respectively, yields

$$f(m_1) \cdot (x_1 - x_0) \leq f(c_1) \cdot (x_1 - x_0) \quad \leq f(M_1) \cdot (x_1 - x_0)$$
$$f(m_2) \cdot (x_2 - x_1) \leq f(c_2) \cdot (x_2 - x_1) \quad \leq f(M_2) \cdot (x_2 - x_1)$$
$$\vdots \qquad\qquad \vdots \qquad\qquad \vdots$$
$$f(m_n) \cdot (x_n - x_{n-1}) \leq f(c_n) \cdot (x_n - x_{n-1}) \leq f(M_n) \cdot (x_n - x_{n-1}).$$

By adding we obtain

$$\sum_{i=1}^{i=n} f(m_i) \cdot (x_i - x_{i-1}) \leq \sum_{i=1}^{i=n} f(c_i) \cdot (x_i - x_{i-1}) \leq \sum_{i=1}^{i=n} f(M_i) \cdot (x_i - x_{i-1}).$$

Since

$$\sum_{i=1}^{i=n} f(m_i) \cdot (x_i - x_{i-1}) \to K \qquad \text{and} \qquad \sum_{i=1}^{i=n} f(M_i) \cdot (x_i - x_{i-1}) \to K$$

we have that

$$\sum_{i=1}^{i=n} f(c_i) \cdot (x_i - x_{i-1}) \to K$$

since this sum is trapped between the lower-bound and upper-bound sums. Thus we have shown that

$$\lim_{\substack{n \to \infty \\ \Delta x_i \to 0}} \sum_{i=1}^{i=n} f(c_i) \cdot (x_i - x_{i-1}) = K.$$

31. Improper integrals

From our discussion of integration we have seen that

$$\int_1^t \frac{dx}{x^2} = \int_1^t x^{-2}\, dx = -1x^{-1}\Big]_1^t = -\frac{1}{x}\Big]_1^t = -\frac{1}{t} + 1.$$

Let us further observe that

$$\lim_{t \to \infty} \int_1^t \frac{dx}{x^2} = \lim_{t \to \infty}\left(-\frac{1}{t} + 1\right) = 1.$$

This sequence of operations defines the improper integral denoted by

$$\int_1^\infty \frac{dx}{x^2}$$

which is said to have value 1.

Geometrically speaking, the region S bounded above by the graph of $f(x) = 1/x^2$, below by the x-axis, and on the left by the line $x = 1$ is said to have area 1 (see

Figure 35). *S* is an unbounded region since it is not bounded on the right, and the concept of improper integral that we have introduced provides us with a means for extending, in a natural way, the definition of area to regions like *S*.

Let us also note that the "proper" integral

$$\int_1^t \frac{dx}{x^2} = -\frac{1}{t} + 1$$

defines the area of the region S_t bounded above by the graph of $f(x) = 1/x^2$, below by the *x*-axis, on the left by the line $x = 1$, and on the right by the line $x = t$ (see Figure 36). Thus, geometrically speaking, the area of *S*, defined by the improper integral $\int_1^\infty \frac{dx}{x^2}$, is the limit as $t \to \infty$ of the area of S_t, defined by the "proper" integral $\int_1^t \frac{dx}{x^2}$.

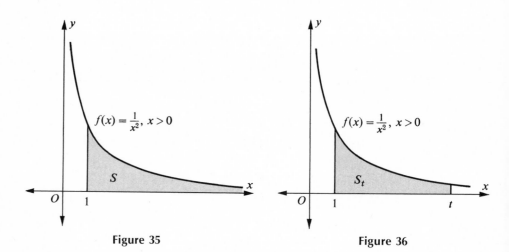

Figure 35   **Figure 36**

More generally, the *improper integrals*

$$\int_a^\infty f(x)\,dx, \qquad \int_{-\infty}^a f(x)\,dx, \qquad \text{and} \qquad \int_{-\infty}^\infty f(x)\,dx$$

are defined as follows.

1. $\displaystyle\int_a^\infty f(x)\,dx = \lim_{t\to\infty}\int_a^t f(x)\,dx$, if this limit exists.

2. $\displaystyle\int_{-\infty}^a f(x)\,dx = \lim_{t\to-\infty}\int_t^a f(x)\,dx$, if this limit exists.

3. $\displaystyle\int_{-\infty}^{\infty} f(x)\,dx = \int_{-\infty}^{0} f(x)\,dx + \int_{0}^{\infty} f(x)\,dx$, if $\displaystyle\int_{-\infty}^{0} f(x)\,dx = \lim_{t\to-\infty}\int_{t}^{0} f(x)\,dx$

and $\displaystyle\int_{0}^{\infty} f(x)\,dx = \lim_{t\to\infty}\int_{0}^{t} f(x)\,dx$ and both these limits exist.

In definition 1 it is assumed that $y = f(x)$ is continuous for $x \ge a$; in definition 2 it is assumed that $y = f(x)$ is continuous for $x \le a$; and in definition 3 it is assumed that $y = f(x)$ is continuous for all x.

EXAMPLE 21. Find, if it exists, the value of $\displaystyle\int_{-\infty}^{-1}\frac{dx}{x^2}$.

Solution. Since

$$\int_{t}^{-1}\frac{dx}{x^2} = -1x^{-1}\Big]_{t}^{-1} = -\frac{1}{x}\Big]_{t}^{-1} = 1 + \frac{1}{t}$$

we have the following.

$$\int_{-\infty}^{-1}\frac{dx}{x^2} = \lim_{t\to-\infty}\int_{t}^{-1}\frac{dx}{x^2} = \lim_{t\to-\infty}\left(1 + \frac{1}{t}\right) = 1$$

EXAMPLE 22. Find, if it exists, the value of $\displaystyle\int_{1}^{\infty}\frac{dx}{x}$.

Solution. Since

$$\int_{1}^{t}\frac{dx}{x} = \ln t - \ln 1 = \ln t - 0 = \ln t$$

we have the following.

$$\int_{1}^{\infty}\frac{dx}{x} = \lim_{t\to\infty}\int_{1}^{t}\frac{dx}{x} = \lim_{t\to\infty}\ln t = \infty$$

Thus the improper integral $\displaystyle\int_{1}^{\infty}\frac{dx}{x}$ does not exist.

A perpetual income stream situation

In Section 29 we saw that if income is obtained continuously over time at a rate of $f(t)$ dollars per year t years from the present ($t = 0$), where the interest rate is r per annum compounded continuously, then the present value Π of the income stream from the present ($t = 0$) to time $t = x$ years is defined by

$$\Pi = \int_{0}^{x} f(t)e^{-rt}\,dt.$$

If the income flow were to persist indefinitely, as in the case of interest from a perpetual bond or revenue from an indestructible asset such as land, then the present value Π of the income stream is defined by the improper integral

$$\Pi = \int_0^\infty f(t)e^{-rt}\, dt.$$

To illustrate, consider a perpetual income stream flowing at the constant rate of $1000 per year, where the interest rate is 10% per annum compounded continuously. The present value Π of this income stream is

$$\Pi = \int_0^\infty 1000\, e^{-0.1t}\, dt.$$

Now

$$\int_0^t 1000\, e^{-0.1t}\, dt = \left. (-10,000\, e^{-0.1t}) \right]_0^t$$

$$= -10,000\, e^{-0.1t} - (-10,000\, e^0)$$
$$= -10,000\, e^{-0.1t} + 10,000.$$

Thus

$$\lim_{t\to\infty} \int_0^t 1000\, e^{-0.1t}\, dt = \lim_{t\to\infty} (-10,000\, e^{-0.1t} + 10,000) = 10,000$$

since

$$e^{-0.1t} = \frac{1}{e^{0.1t}} \to 0 \text{ as } t \to \infty.$$

Therefore

$$\Pi = \int_0^\infty 1000\, e^{-0.1t}\, dt = \$10,000.$$

Exercises

Find the value of each of the following improper integrals, if it exists.

141. $\int_1^\infty \dfrac{dx}{x^3}$

142. $\int_{-\infty}^{-1} \dfrac{dx}{x^4}$

143. $\int_4^\infty \dfrac{dx}{\sqrt{x}}$

144. $\int_0^\infty e^{-x}\, dx$

145. $\int_0^\infty xe^{-x^2}\, dx$

146. $\int_{-\infty}^3 \dfrac{dx}{(4-x)^2}$

147. $\int_0^\infty \dfrac{x}{1+x^2}\, dx$

148. $\int_{-\infty}^0 e^x\, dx$

149. $\displaystyle\int_{1}^{\infty} \frac{dx}{x\sqrt{x}}$ 150. $\displaystyle\int_{-\infty}^{3} \frac{dx}{\sqrt{9-x}}$

151. Find the present value of a perpetual income stream flowing at the constant rate of $4000 per year, where the interest rate is 10% per annum compounded continuously.

152. Find the present value of a perpetual income stream flowing at the constant rate of $1200 per year, where the interest rate is 12% per annum compounded continuously.

153. Show that the present value of a perpetual income stream flowing at the constant rate of A dollars per year, where the interest rate is r per annum compounded continuously, is A/r dollars.

6

Topics in multivariable calculus

32. Multivariable functions

Thus far we have restricted our attention to functions involving two variables, x and y, let us say, and a rule or relation by means of which there is assigned to each value of x exactly one value of y. Such functions are called functions of one independent variable; y is said to be a function of the single variable x.

In many situations, several quantities are interacting, and the behavior of one in terms of another can only be studied by the simplification achieved using the sometimes unrealistic assumption that all other factors are held constant. Thus, for example, the market demand for a commodity depends not only on the price of the commodity but also on such factors as prices of related commodities, taste, and income. The total profit of a firm manufacturing several products depends on the amounts produced and sold, and costs. The capital accumulated from an investment depends on the amount initially invested, interest conditions, and time.

The dependence of one quantity on a number of other quantities can often be described by "functions of more than one variable," or *multivariable functions*, as they are sometimes called. A function f is said to be a *function of x and y* if to each ordered pair of values (x,y) there is assigned exactly one value. The *domain of definition of f* consists of all ordered pairs of values (x,y) for which its rule is defined. If no explicit mention is made of the domain of definition of f, it is understood to consist of all ordered pairs of numbers for which the rule of f is meaningful.

The symbol $f(x,y)$ is used to denote the function f, the rule of function f, as well as the value assigned by f to (x,y). This overuse of one symbol rarely leads to confusion since the context of its use usually makes clear the aspect of the function which is of concern. Sometimes it is convenient to use a single letter to denote the value assigned by f to (x,y). If u is used, then u and $f(x,y)$ denote the same quantity and we thus write $u = f(x,y)$.

The function f defined by

$$f(x,y) = x^2y + 2y$$

or

$$u = x^2y + 2y$$

has as its domain of definition all ordered pairs of real numbers. The values assigned, for example, to $(2, 3)$ and $(3, 2)$ are

$$f(2, 3) = (2)^23 + 2(3) = 18$$
$$f(3, 2) = (3)^22 + 2(2) = 22.$$

Thus $f(2, 3) \neq f(3, 2)$, which serves to emphasize that the *order* of the values in an ordered pair is most significant.

Functions of three, four, or more, independent variables are defined in an analogous way. Thus, for example, a function f is said to be a *function of x, y, and z* if to each ordered triple of values (x, y, z) in its domain of definition there is assigned exactly one value. The *domain of definition of f* consists of all ordered triples of values (x, y, z) for which its rule is defined.

The function f defined by

$$f(x,y,z) = \frac{x^2 + y^2}{z}$$

has all ordered triples (x, y, z), with $z \neq 0$, as its domain of definition. Thus, for example,

$$f(1, 2, 3) = \frac{1^2 + 2^2}{3^2} = \frac{5}{9}.$$

EXAMPLE 1. *A Capital Accumulation Function.* The amount accumulated from an investment depends on the initial amount invested, interest conditions, and time. Specifically, the function

$$f(y,r,t) = ye^{rt}$$

describes the amount accumulated after t years when an initial amount of y dollars is invested at a rate of r per annum compounded continuously (see Section 7 for a discussion of continuous compounding of interest).

EXAMPLE 2. *Demand Functions.* A demand function expresses the amount q of a commodity purchased per unit time in a market as a function of p, the price of the commodity, x, a measure of the prices of related commodities, y, a measure of the incomes of buyers, and t, a measure of the tastes of buyers. A demand function thus has the general form

$$q = f(p, x, y, t).$$

If income, the prices of related commodities, and taste are all assumed to be constant, then q becomes a function of one variable p representing the price of the commodity,

$$q = h(p).$$

This important special case was discussed in Example 5, Section 1.

One specific function type that has been used in the study of demand is the linear form

$$q = a + bp + cx + dy$$

where a, b, c, and d are constants. In his now classic study of demand functions,* Henry Schultz obtained

$$q = 3.4892 - 0.0899p + 0.0637x + 0.0187y$$

as a demand function for beef. The related commodity in this situation is pork.

Another function type that has been used in the study of demand is the nonlinear form

$$q = ap^b x^c y^d t^f$$

where a, b, c, d, and f are constants. In his paper "The Analysis of Market Demand,"† R. Stone obtained the function

$$q = 1.058p^{0.727} x^{0.914} y^{0.136} t^{0.816}$$

in his analysis of the demand for beer in the United Kingdom. x represents the average retail price of all other related commodities and t is an index of the strength of beer which can be interpreted as an index of taste.

EXAMPLE 3.   *Production Functions.* A production function expresses the output of a firm or industry per unit time as a function of the inputs utilized in the production process. If, for example, q is the output when amounts x, y, z, w of four productive factors are employed, then the production function can be expressed in function notation by

$$q = f(x, y, z, w).$$

To illustrate, suppose that

$$q = 6xy - 3x^2 - y^2$$

bushels of rye are produced per week when x man-hours of labor are employed on y acres of land. If 300 acres are cultivated and 500 man-hours of labor are employed, then the weekly output is

---

*The Theory and Measurement of Demand (Chicago: University of Chicago Press, 1938).
†Journal of the Royal Statistical Society, vol. 108 (1945).

$$q = 6(500)(300) - 3(500)^2 - (300)^2 = 60{,}000 \text{ bushels.}$$

One widely used production function form in economic analysis, whose use was pioneered by former Senator Paul H. Douglas and his collaborators, is the Cobb-Douglas function,

$$q = ax^b y^{1-b}.$$

Here q is output, x is a measure of the quantity of labor employed, y is a measure of the quantity of capital employed, and a and b $(0 < b < 1)$ are constants. Douglas and his associates derived production functions for whole economies and sectors of economies. In a 1928 paper,* Douglas and C. W. Cobb employed the production function

$$q = 1.01 x^{0.75} y^{0.25}$$

in a study of manufacturing in the United States.

Exercises

1. For $f(x,y) = 2xy - x^2 y^2$, find $f(1,3)$, $f(-1,2)$, $f(3,1)$, and $f(2,6)$.
2. For $f(x,y,z) = 2xy + y^2 z$, find $f(2,3,1)$, $f(-1,2,4)$, and $f(3,2,-1)$.
3. For $f(x,y,z) = \dfrac{x^2 + y^2}{z^2}$, find $f(2,3,1)$, $f(-1,2,4)$, and $f(3,2,1)$.

33. Partial derivatives

If all the independent variables of a multivariable function except one are held constant, then the multivariable function simplifies to a function of that one variable. To illustrate, consider

$$f(x,y,z) = 2x^2 + y^2 z$$

or

$$u = 2x^2 + y^2 z.$$

If $y = 2$ and $z = 1$, then we obtain a function of x,

$$g(x) = f(x, 2, 1) = 2x^2 + 4.$$

Since this is a function of one variable, derivative questions can be entertained.

$$g'(x) = 4x$$

exists for all values of x. For $x = 1$ we have

$$g'(1) = 4.$$

For $x = 3$ we have

$$g'(3) = 12.$$

---

*"A Theory of Production," *American Economic Review*, vol. 18 (1928).

The derivative of $g(x) = f(x, 2, 1) = 2x^2 + 4$ at $x = 1$ is called the partial derivative of function f with respect to x at the point $(x, y, z) = (1, 2, 1)$ and can be denoted by any of the following notations.

$$\frac{\partial f}{\partial x}\Big|_{\substack{x=1 \\ y=2 \\ z=1}} \qquad \frac{\partial f}{\partial x}\Big|_{\substack{1 \\ 2 \\ 1}} \qquad \frac{\partial u}{\partial x}\Big|_{\substack{1 \\ 2 \\ 1}} \qquad f_x(1, 2, 1) \qquad u_x(1, 2, 1)$$

Thus we have

$$\frac{\partial f}{\partial x}\Big|_{\substack{1 \\ 2 \\ 1}} = 4 \quad \text{or} \quad f_x(1, 2, 1) = 4$$

and so on.

The derivative of $g(x) = f(x, 2, 1) = 2x^2 + 4$ at $x = 3$ is called the partial derivative of f with respect to x at $(3, 2, 1)$. Thus

$$\frac{\partial f}{\partial x}\Big|_{\substack{3 \\ 2 \\ 1}} = 12 \quad \text{or} \quad f_x(3, 2, 1) = 12$$

and so on.

If x and y are held constant, then $f(x, y, z)$ simplifies to a function of z. For $x = 2$ and $y = 1$, for example, we obtain

$$f(2, 1, z) = z + 8.$$

The derivative of this function at $z = 3$, called the partial derivative of f with respect to z at $(2, 1, 3)$, and denoted by any of the following notations,

$$\frac{\partial f}{\partial z}\Big|_{\substack{2 \\ 1 \\ 3}} \qquad f_z(2, 1, 3) \qquad \frac{\partial u}{\partial z}\Big|_{\substack{2 \\ 1 \\ 3}} \qquad u_z(2, 1, 3),$$

is

$$\frac{\partial f}{\partial z}\Big|_{\substack{2 \\ 1 \\ 3}} = 1 \quad \text{or} \quad f_z(2, 1, 3) = 1.$$

If x and z are held constant, then $f(x, y, z)$ simplifies to a function of y. For $x = 1$ and $z = 3$, we obtain

$$f(1, y, 3) = 3y^2 + 2.$$

The derivative of this function at $y = -1$, called the partial derivative of f with respect to y at $(1, -1, 3)$, and denoted by any of the following notations,

$$\frac{\partial f}{\partial y}\Big|_{\substack{1 \\ -1 \\ 3}} \qquad f_y(1, -1, 3) \qquad \frac{\partial u}{\partial y}\Big|_{\substack{1 \\ -1 \\ 3}} \qquad u_y(1, -1, 3)$$

is

$$\frac{\partial f}{\partial y}\bigg|_{\substack{1\\-1\\3}} = -6 \quad \text{or} \quad f_y(1,-1,3) = -6.$$

Thus the multivariable counterpart of the derivative of a function of one variable is the concept of partial derivative.

If f is a multivariable function, one of whose independent variables is x, and the other independent variables are held constant, then f simplifies to a function of x. If the derivative of this function of x exists, then it is called the *partial derivative of f with respect to x* and is denoted by

$$\frac{\partial f}{\partial x} \quad \text{or} \quad f_x.$$

Partial derivatives of f with respect to its other independent variables are defined in a completely analogous way.

EXAMPLE 4.   For $f(x,y) = 2x^2y^3$, determine $\dfrac{\partial f}{\partial x}$ and $\dfrac{\partial f}{\partial y}$.

Solution.   If we consider y as being held constant, then $f(x,y) = 2x^2y^3$ simplifies to a function of x which is the constant $2y^3$ times the function x^2. Differentiating with respect to x, using the theorem that the derivative of a constant $(2y^3)$ times a function (x^2) is the constant $(2y^3)$ times the derivative of the function (x^2), yields

$$\frac{\partial f}{\partial x} = (2y^3) \cdot \frac{d(x^2)}{dx} = 2y^3(2x) = 4y^3x.$$

If we consider x as being held constant, then $f(x,y) = 2x^2y^3$ simplifies to a function of y which is the constant $2x^2$ times the function y^3. Differentiating with respect to y yields

$$\frac{\partial f}{\partial y} = (2x^2) \cdot \frac{d(y^3)}{dy} = (2x^2)(3y^2) = 6x^2y^2.$$

EXAMPLE 5.   For $f(x,y,z) = \dfrac{x^2z^2}{2xy + z}$, determine $\dfrac{\partial f}{\partial z}$.

Solution.   If we consider x and y as being held constant, then $f(x,y,z) = (x^2z^2)/(2xy + z)$ simplifies to a function of z. Differentiating this function of z by means of the quotient theorem and treating x^2 and $2xy$ as constants yields the following.

$$\frac{\partial f}{\partial z} = \frac{(2xy + z) \cdot \dfrac{d(x^2z^2)}{dz} - x^2z^2 \cdot \dfrac{d(2xy + z)}{dz}}{(2xy + z)^2}$$

$$= \frac{(2xy + z)2x^2z - x^2z^2(0 + 1)}{(2xy + z)^2}$$

$$= \frac{4x^3yz + 2x^2z^2 - x^2z^2}{(2xy + z)^2} = \frac{4x^3yz + x^2z^2}{(2xy + z)^2}$$

EXAMPLE 6.   For $f(x,y,z) = e^{xyz}$, determine $\dfrac{\partial f}{\partial x}$.

Solution.   Considering y and z as being held constant and differentiating with respect to x yields

$$\frac{\partial f}{\partial x} = \frac{d(e^{xyz})}{dx}.$$

If we let $u = xyz$, then by the chain rule

$$\frac{d(e^{xyz})}{dx} = \frac{d(e^u)}{du} \cdot \frac{du}{dx}.$$

Since

$$\frac{d(e^u)}{du} = e^u \quad \text{and} \quad \frac{du}{dx} = (yz) \cdot \frac{d(x)}{dx} = yz$$

then

$$\frac{\partial f}{\partial x} = \frac{d(e^{xyz})}{dx} = e^u \cdot (yz)$$

$$= e^{xyz} \cdot (yz) = yze^{xyz}.$$

Exercises

4. For $f(x,y) = 3x^2 + xy^2$, find the following.
   a. The function of x obtained when $y = 1$ and $f_x(3, 1)$.
   b. The function of x obtained when $y = 3$ and $f_x(2, 3)$.
   c. The function of y obtained when $x = 2$ and $f_y(2, 4)$.
   d. The function of y obtained when $x = -1$ and $f_y(-1, 3)$.
5. For $f(x,y,z) = 2xy + yz^2 - x^2z^3$, find the following.
   a. The function of x obtained when $y = 1$, $z = 2$; $f_x(3, 1, 2)$ and $f_x(-1, 1, 2)$.
   b. The function of x obtained when $y = -2$, $z = 2$; $f_x(-1, -2, 2)$ and $f_x(4, -2, 2)$.
   c. The function of y obtained when $x = -2$, $z = 1$; $f_y(-2, 1, 1)$ and $f_y(-2, -1, 1)$.
   d. The function of y obtained when $x = 2$, $z = -2$; $f_y(2, 3, -2)$ and $f_y(2, 0, -2)$.
   e. The function of z obtained when $x = 1$, $y = -1$; $f_z(1, -1, 3)$ and $f_z(1, -1, 4)$.
   f. The function of z obtained when $x = -2$, $y = 1$; $f_z(-2, 1, 2)$ and $f_z(-2, 1, 3)$.
6. For $f(x,y) = 3x^2 + xy^2$, find $\dfrac{\partial f}{\partial x}$ and $\dfrac{\partial f}{\partial y}$.
7. For $f(x,y,z) = 2xy + yz^2 - x^2z^3$, find $\dfrac{\partial f}{\partial x}$, $\dfrac{\partial f}{\partial y}$, and $\dfrac{\partial f}{\partial z}$.
8. For $f(x,y) = 4x^3y^2 + 3x^2 + xy^2$, find $\dfrac{\partial f}{\partial x}$, $\dfrac{\partial f}{\partial y}$, $f_x(1, 2)$, and $f_y(2, -1)$.

9. For $u = 3x^2y + y^2z^3$, find $u_x(x,y,z)$, $u_y(x,y,z)$, $u_z(x,y,z)$, $u_x(1,2,-1)$, $u_y(-1,2,-1)$, and $u_z(2,3,1)$.

10. For $f(x,y,z) = (xy)e^z$, find $\dfrac{\partial f}{\partial x}$, $\dfrac{\partial f}{\partial y}$, and $\dfrac{\partial f}{\partial z}$.

11. For $f(x,y,z) = \dfrac{xy^2 + yz^2}{xyz}$, find $\dfrac{\partial f}{\partial x}$, $\dfrac{\partial f}{\partial y}$, and $\dfrac{\partial f}{\partial z}$.

12. For $f(x,y,z) = \dfrac{2xyz^3}{3x + 2y + 4z}$, find $\dfrac{\partial f}{\partial x}$, $\dfrac{\partial f}{\partial y}$, and $\dfrac{\partial f}{\partial z}$.

13. For $f(x,y) = y^2 \ln x$, find $\dfrac{\partial f}{\partial x}$ and $\dfrac{\partial f}{\partial y}$.

14. For $f(x,y) = \sqrt{x^2 + y^2}$, find $\dfrac{\partial f}{\partial x}$ and $\dfrac{\partial f}{\partial y}$.

15. For $f(x,y,z) = \sqrt{x^2 + 3xy^2z^2}$, find $\dfrac{\partial f}{\partial x}$, $\dfrac{\partial f}{\partial y}$, and $\dfrac{\partial f}{\partial z}$.

16. For $f(x,y,z) = \dfrac{x^2z^2}{2xy + z}$, find $\dfrac{\partial f}{\partial x}$ and $\dfrac{\partial f}{\partial y}$.

17. For $f(x,y,z) = e^{xyz}$, find $\dfrac{\partial f}{\partial y}$ and $\dfrac{\partial f}{\partial z}$.

34. Concepts with a partial derivative structure

The following are a few of many concepts which are defined in terms of partial derivatives.

Marginal productivity

If $q = f(x_1, x_2, \ldots, x_n)$ is a production function which expresses output q as a function of the productive factors $x_1, x_2, \ldots, x_n$, then $\dfrac{\partial q}{\partial x_1}$ is defined as the *marginal productivity of factor x_1* (or *marginal product of x_1*), $\dfrac{\partial q}{\partial x_2}$ is defined as the *marginal productivity of factor x_2*, and so on.

To illustrate, consider the Cobb-Douglas production function,

$$q = 1.01x^{0.75}y^{0.25}$$

used in a study of manufacturing in the United States (see Example 3, Section 32), where x is a measure of quantity of labor and y is a measure of quantity of capital. The marginal productivity of labor is

$$\frac{\partial q}{\partial x} = 1.01y^{0.25} \cdot \frac{d(x^{0.75})}{dx} = 0.7575x^{-0.25}y^{0.25}.$$

The marginal productivity of capital is

$$\frac{\partial q}{\partial y} = 1.01x^{0.75} \cdot \frac{d(y^{0.25})}{dy} = 0.2525x^{0.75}y^{-0.75}.$$

Price elasticity of demand

Suppose that the demand functions for n related commodities $A_1, A_2, \ldots, A_n$ in terms of their respective prices $p_1, p_2, \ldots, p_n$ and income y are

$$q_1 = f_1(p_1, p_2, \ldots, p_n, y)$$
$$q_2 = f_2(p_1, p_2, \ldots, p_n, y)$$
$$\vdots$$
$$q_n = f_n(p_1, p_2, \ldots, p_n, y).$$

Then we have the following definitions.

$$\eta_{11} = -\frac{p_1}{q_1} \cdot \frac{\partial q_1}{\partial p_1}$$

is called the *partial elasticity of demand for commodity A_1 with respect to its price p_1.*

$$\eta_{22} = -\frac{p_2}{q_2} \cdot \frac{\partial q_2}{\partial p_2}$$

is called *the partial elasticity of demand for commodity A_2 with respect to its price p_2,* and so on. These concepts are direct extensions of ordinary demand elasticity in which demand is regarded as a function of only the price of the good concerned.

Consider the demand function obtained by Henry Schultz for beef (see Example 2, Section 32).

$$q_1 = 3.4892 - 0.0899p_1 + 0.0637p_2 + 0.0187y$$

where p_1 is the price of beef, p_2 is the price of pork, and y is a measure of income. We obtain

$$\eta_{11} = -\frac{p_1}{q_1} \cdot \frac{\partial q_1}{\partial p_1} = -\frac{p_1}{q_1}(-0.0899) = \frac{(0.0899)p_1}{q_1}$$

as the partial elasticity of demand for beef with respect to its price.

Price cross-elasticity of demand

Again, suppose that the demand functions for n related commodities $A_1, A_2, \ldots, A_n$, in terms of their respective prices $p_1, p_2, \ldots, p_n$ and income y are

$$q_1 = f_1(p_1, p_2, \ldots, p_n, y)$$
$$q_2 = f_2(p_1, p_2, \ldots, p_n, y)$$
$$\vdots$$
$$q_n = f_n(p_1, p_2, \ldots, p_n, y).$$

Partial elasticities of demand for one commodity with respect to the price of another commodity are defined as follows.

$$\eta_{12} = \frac{p_2}{q_1} \cdot \frac{\partial q_1}{\partial p_2} \qquad \text{The partial elasticity of demand of } A_1 \text{ with respect to the price of } A_2.$$

$$\eta_{13} = \frac{p_3}{q_1} \cdot \frac{\partial q_1}{\partial p_3} \qquad \text{The partial elasticity of demand of } A_1 \text{ with respect to the price of } A_3.$$

$$\eta_{21} = \frac{p_1}{q_2} \cdot \frac{\partial q_2}{\partial p_1} \qquad \text{The partial elasticity of demand of } A_2 \text{ with respect to the price of } A_1.$$

More generally, the *partial elasticity of demand of good A_i with respect to the price of good A_j* is defined by

$$\eta_{ij} = \frac{p_j}{q_i} \cdot \frac{\partial q_i}{\partial p_j}.$$

From Henry Schultz's demand function for beef,

$$q_1 = 3.4892 - 0.0899p_1 + 0.0637p_2 + 0.0187y$$

we obtain

$$\eta_{12} = \frac{p_2}{q_1} \cdot \frac{\partial q_1}{\partial p_2} = \frac{p_2}{q_1}(0.0637) = \frac{0.0637p_2}{q_1}$$

as the partial elasticity of demand of beef with respect to the price of pork.

Income elasticity of demand

Once again, suppose that the demand functions for n related quantities $A_1, A_2, \ldots, A_n$ in terms of their respective prices $p_1, p_2, \ldots, p_n$ and income y are

$$q_1 = f_1(p_1, p_2, \ldots, p_n, y)$$
$$q_2 = f_2(p_1, p_2, \ldots, p_n, y)$$
$$\vdots$$
$$q_n = f_n(p_1, p_2, \ldots, p_n, y).$$

Then the *income elasticity of demand for commodity A_1* is defined as

$$\eta_{1y} = \frac{y}{q_1} \cdot \frac{\partial q_1}{\partial y}.$$

The *income elasticity of demand for commodity A_2* is defined as

$$\eta_{2y} = \frac{y}{q_2} \cdot \frac{\partial q_2}{\partial y}$$

and so on.

From Henry Schultz's demand function for beef,

$$q_1 = 3.4892 - 0.0899p_1 + 0.0637p_2 + 0.0187y$$

we obtain

$$\eta_{1y} = \frac{y}{q_1} \cdot \frac{\partial q_1}{\partial y} = \frac{y}{q_1} \cdot (0.0187) = \frac{0.0187y}{q_1}$$

as the income elasticity of demand for beef.*

35. Higher order partial derivatives

The partial derivatives of $u = f(x,y)$ are also functions of x and y and thus they themselves may have partial derivatives. If they exist, these partial derivatives are called the *second partial derivatives* of $u = f(x,y)$ and are denoted as follows.

$$\frac{\partial}{\partial x}\left(\frac{\partial u}{\partial x}\right) = \frac{\partial^2 u}{\partial x^2} = u_{xx}(x,y) = \frac{\partial^2 f}{\partial x^2} = f_{xx}(x,y)$$

$$\frac{\partial}{\partial y}\left(\frac{\partial u}{\partial y}\right) = \frac{\partial^2 u}{\partial y^2} = u_{yy}(x,y) = \frac{\partial^2 f}{\partial y^2} = f_{yy}(x,y)$$

$$\frac{\partial}{\partial x}\left(\frac{\partial u}{\partial y}\right) = \frac{\partial^2 u}{\partial x \partial y} = u_{yx}(x,y) = \frac{\partial^2 f}{\partial x \partial y} = f_{yx}(x,y)$$

$$\frac{\partial}{\partial y}\left(\frac{\partial u}{\partial x}\right) = \frac{\partial^2 u}{\partial y \partial x} = u_{xy}(x,y) = \frac{\partial^2 f}{\partial y \partial x} = f_{xy}(x,y)$$

Note that the subscripts are reversed when taking higher order partial derivatives in one notation, but not in the other. For example,

$$\frac{\partial}{\partial y}\left(\frac{\partial u}{\partial x}\right) = u_{xy}(x,y) = \frac{\partial^2 u}{\partial y \partial x}.$$

EXAMPLE 7. For $f(x,y) = 2x^2y^3$, then

$$\frac{\partial f}{\partial x} = 4xy^3 \qquad \frac{\partial^2 f}{\partial x^2} = 4y^3 \qquad \frac{\partial^2 f}{\partial y \partial x} = \frac{\partial(4xy^3)}{\partial y} = 12xy^2$$

$$\frac{\partial f}{\partial y} = 6x^2y^2 \qquad \frac{\partial^2 f}{\partial y^2} = 12x^2y \qquad \frac{\partial^2 f}{\partial x \partial y} = \frac{\partial(6x^2y^2)}{\partial x} = 12xy^2.$$

From Example 7 we see that if $f(x,y) = 2x^2y^3$, then

$$\frac{\partial^2 f}{\partial y \partial x} = \frac{\partial^2 f}{\partial x \partial y} = 12xy^2.$$

---

*For further discussion of these concepts, see, for example, C. E. Ferguson, *Microeconomic Theory*, rev. ed. (Homewood, Illinois: Richard D. Irwin, Inc., 1969), Chapter 4; R. G. D. Allen, *Mathematical Analysis for Economists* (New York: Macmillan Company, 1938), Section 12.6.

This is not accidental. It can be shown that if the partial derivatives of $u = f(x,y)$ satisfy certain conditions, then

$$\frac{\partial^2 f}{\partial y \partial x} = \frac{\partial^2 f}{\partial x \partial y}.$$

Third and higher order partial derivatives of $u = f(x,y)$ are defined in a similar way.

$$\frac{\partial}{\partial x}\left(\frac{\partial^2 u}{\partial x^2}\right) = \frac{\partial^3 u}{\partial x^3}$$

$$\frac{\partial}{\partial x}\left(\frac{\partial^2 u}{\partial x \partial y}\right) = \frac{\partial^3 u}{\partial x^2 \partial y}$$

$$\frac{\partial}{\partial x}\left(\frac{\partial^2 u}{\partial y \partial x}\right) = \frac{\partial^3 u}{\partial x \partial y \partial x}$$

And so on.

For $f(x,y) = 2x^2 y^3$, then

$$\frac{\partial^3 f}{\partial x^3} = \frac{\partial (4y^3)}{\partial x} = 0$$

$$\frac{\partial^3 f}{\partial x^2 \partial y} = \frac{\partial (12xy^2)}{\partial x} = 12y^2.$$

Similar notation is used for higher order partial derivatives of functions of three or more variables.

Exercises

18. Find all second partial derivatives of $f(x,y) = 3x^3 + xy^3$ and $f_{xx}(1,2), f_{yy}(-1,3)$, $f_{xy}(2,3)$, and $f_{yx}(4,1)$.
19. Find all second partial derivatives of $f(x,y) = 4x^3 y^4$ and $f_{xx}(1,2)$, $f_{yy}(3,2)$, $f_{xy}(3,2)$, and $f_{yx}(-1,-2)$.
20. Find all second partial derivatives of $f(x,y) = 2x^3 y^2 - y^4 x^2$ and $f_{xx}(-1,2)$, $f_{yy}(1,3)$, $f_{xy}(2,3)$, and $f_{yx}(4,1)$.
21. Find all second partial derivatives of $f(x,y,z) = 4x^3 y^2 z^4 + xyz$.
22. Find all second partial derivatives of $f(x,y,z) = 4x^5 y^3 + x^3 z^4$.

36. Multivariable optimization problems

Many problems lead one to seek the greatest or least value of a multivariable function, and in this section we provide an introduction to some of the fundamental results of multivariable optimization theory. While our discussion is restricted to functions of two independent variables, much of what is said can be extended to functions of three or more variables in a natural way.

The definitions of absolute and local maximum and minimum values for multivariable functions are analogous to those given for their single variable cousins. A function f of x and y is said to have an *absolute maximum value* at (a, b) if $f(a, b) \geq f(x, y)$ for all (x, y) in the domain of definition of f. $f(a, b)$ itself is said to be the (*absolute*) *maximum value* of f. $f(x, y) = -x^2 - y^2$, for example, clearly has an absolute maximum value at $(0, 0)$.

Function f is said to have a *local maximum value* at (a, b) if there is a circle C centered at (a, b) such that $f(a, b) \geq f(x, y)$ for all (x, y) in C which are also in the domain of definition of f. $f(a, b)$ is said to be a *local maximum value* of f.

The concepts of *absolute minimum value* and *local minimum value* are defined in a completely analogous way.

The maximum and minimum values of a multivariable function f are called its *extreme values*. As is the case for functions of one variable, the search for extreme values of a multivariable function f begins with a search for its *critical* points—points in the domain of definition of f which can give rise to extreme values. Critical points for a multivariable function f include points on boundary curves or lines of the domain of definition of f (the counterpart of end points of the domain of definition in the one-variable case) and points at which the partial derivatives of f are zero (the counterpart of values at which the derivative is zero in the one-variable case). For the latter of these situations, the following theorem holds.

Theorem. Suppose f, defined on a region R, has an extreme value (local or absolute) at the point (a, b) of R, and (a, b) is interior to some circle C centered at (a, b) which is in R (that is, (a, b) is not on the boundary of R). If f and its first partial derivatives satisfy *certain conditions* within C, then

$$f_x(a, b) = 0 \qquad f_y(a, b) = 0.$$

Thus to find critical points of this type, determine the partial derivatives of f, set them equal to zero, and solve for x and y. To illustrate, consider the problem of finding the extreme values of

$$f(x, y) = -x^2 - y^2.$$

The first partial derivatives of f are

$$f_x(x, y) = -2x$$
$$\text{and}$$
$$f_y(x, y) = -2y.$$

Setting these partial derivatives equal to zero yields

$$-2x = 0$$
$$-2y = 0$$

from which we obtain

$$x = 0$$
$$y = 0.$$

Thus $(0, 0)$ is a critical point of $f(x,y) = -x^2 - y^2$. From the nature of this function it is immediately clear that $f(0,0) = 0$ is an absolute maximum value.

The general situation is, however, more complicated. As in the one-variable case, not every critical point need necessarily yield an extreme value. Therefore, means must be found to distinguish those critical points which yield extreme values from those which do not. The following theorem is a multivariable counterpart of the second derivative test in the one-variable situation.

Theorem. Let f denote a function of x and y which is defined and "well behaved" in some circular region centered at the critical point (a, b) $(f_x(a, b) = 0, f_y(a, b) = 0)$. Consider

$$F(x,y) = \left(\frac{\partial^2 f}{\partial x^2}\right)\left(\frac{\partial^2 f}{\partial y^2}\right) - \left(\frac{\partial^2 f}{\partial x \partial y}\right)^2.$$

1. If $F(a, b) > 0$ and $f_{xx}(a, b) > 0$, then $f(a, b)$ is a local minimum value.
2. If $F(a, b) > 0$ and $f_{xx}(a, b) < 0$, then $f(a, b)$ is a local maximum value.
3. If $F(a, b) < 0$, then $f(a, b)$ is not an extreme value.
4. If $F(a, b) = 0$, then the test fails to yield any information and no conclusion can be drawn.

By f "well behaved" in some circular region centered at the critical point (a, b), we mean that f must satisfy certain rather technical and complicated conditions in order for the test to be applicable. These conditions are satisfied by most functions met in practice and are satisfied by all functions considered in this section. Thus we will not go into the definition of these conditions here.

To illustrate, let us again consider

$$f(x,y) = -x^2 - y^2$$

which, as we have observed, has a critical point $(0, 0)$ at which the maximum value $f(0, 0) = 0$ is assumed. Since

$$\frac{\partial f}{\partial x} = -2x \quad \text{and} \quad \frac{\partial f}{\partial y} = -2y$$

we obtain

$$\frac{\partial^2 f}{\partial x^2} = -2 \qquad \frac{\partial^2 f}{\partial y^2} = -2 \qquad \frac{\partial^2 f}{\partial x \partial y} = 0.$$

Thus

$$F(x,y) = (-2)(-2) - (0)^2 = 4$$

and it follows that

$$F(0, 0) = 4.$$

Since $F(0, 0) = 4$, then $F(0, 0) > 0$, and since $f_{xx}(0, 0) = -2$, then $f_{xx}(0, 0) < 0$. Thus it follows from part 2 of our theorem that $f(0, 0) = 0$ is a local maximum value. From the nature of $f(x,y) = -x^2 - y^2$, we see that $f(0, 0) = 0$ is an absolute maximum value.

EXAMPLE 8.   A coffee producer who produces two blends of coffee obtains a weekly profit of

$$f(x,y) = 10x + 12y + 4xy - x^2 - 5y^2$$

dollars for an output of x tons of Blend I and y tons of Blend II per week. Determine the output (x,y) for which profit is maximized.

Solution.   To find critical points of f, we determine $f_x(x,y)$ and $f_y(x,y)$, set them equal to zero, and solve for x and y.

$$\frac{\partial f}{\partial x} = 10 + 4y - 2x$$

$$\frac{\partial f}{\partial y} = 12 + 4x - 10y$$

Setting these partial derivatives equal to zero yields

$$10 + 4y - 2x = 0$$
$$12 + 4x - 10y = 0$$

from which we obtain

$$-2x + 4y = -10$$
$$2x - 5y = -6.$$

On solving this system of equations we obtain

$$x = 37$$
$$y = 16.$$

Therefore $(37, 16)$ is a critical point.

$$\frac{\partial^2 f}{\partial x^2} = -2 \qquad \frac{\partial^2 f}{\partial y^2} = -10 \qquad \frac{\partial^2 f}{\partial x \partial y} = 4$$

Thus

$$F(x,y) = (-2)(-10) - (4)^2 = 4$$

and it follows that

$$F(37, 16) = 4.$$

Since $F(37, 16) > 0$, and $f_{xx}(37, 16) = -2$ so that $f_{xx}(37, 16) < 0$, it follows from part 2 of our theorem that the coffee producer's profit function has a local maximum value at $(37, 16)$. This maximum value is

$$f(37, 16) = 10(37) + 12(16) + 4(37)(16) - (37)^2 - 5(16)^2 = 281.$$

The coffee producer should produce 37 tons of Blend I and 16 tons of Blend II per week to obtain a maximum weekly profit of $281.

In many situations of interest a multivariable function assumes an extreme value on the boundary of its domain of definition, and the approach developed in this section is not applicable.*

Exercises

Examine the following functions for extreme values.

23. $f(x,y) = x^2 + y^2$
24. $f(x,y) = 2x^2 - xy + y^2$
25. $f(x,y) = x^2 + 4y^2 - 4x$
26. $f(x,y) = x^3 - 3x - y^2$
27. $f(x,y) = x^2 - y^2$
28. $f(x,y) = 2x^2 - 3xy + y^2$
29. $f(x,y) = x^2 + 2xy + y^2 - 3y + 2x$
30. $f(x,y) = 6x + 4y + 10xy - 2x^2 - 3y^2$
31. $f(x,y) = x^3 - xy^2 + 4x$
32. $f(x,y) = x^3 + xy^2 - 12x$
33. $f(x,y) = x^3 + x^2y^2 - 27x$

34. A manufacturer of shipping containers received an order to make ten thousand rectangular boxes. Each is to have a volume of eight cubic feet. Since his cost of materials is minimized when the surface area of a box is a minimum, the manufacturer is interested in finding the box dimensions for which the volume is eight cubic feet and surface area is minimized.

   a. Show that if the dimensions of the box are x, y, and z, then the surface area of the box is given by the function $S(x,y) = 2\left(xy + \dfrac{8}{x} + \dfrac{8}{y}\right)$, where $xyz = 8$.

   b. Find the box dimensions for which surface area is minimized.

---

*A very important special case of this set of circumstances is considered in Part 1, "Linear Programming," of my *Finite Mathematics for Business and Social Science* (Lexington, Mass.: Xerox College Publishing, 1974).

Appendix 1

The straight line

Let us recall that if $P(x, y)$ and $R(a, b)$ are two points on a nonvertical line L, then the *slope of the line segment determined by P and R* is defined as the ratio of the difference in the y-values of the points to the difference in the x-values taken in the same order. That is,

$$\frac{y - b}{x - a}.$$

Thus the slope of the line segment with end points $P(1, 3)$ and $R(4, 7)$ is

$$\frac{7 - 3}{4 - 1} = \frac{4}{3}.$$

To obtain an equation for line L, some property which sets it apart from all other curves must be given. We will assume as a working definition of the straight line that it is the only curve with the property that the slope of any line segment on L is equal to the slope of any other line segment on L. This common value is called the *slope of line L.* Thus if the slope of L is m and $R(a, b)$ is a point on L, then $P(x, y)$ is on L if and only if

$$\frac{y - b}{x - a} = m. \tag{1}$$

Multiplying both members of (1) by $x - a$ yields

$$y - b = m(x - a)$$

as an equation for L. A point $P(x, y)$ is on L if and only if it satisfies the equation $y - b = m(x - a)$.

EXAMPLE 1. An equation of the line L with slope 8 which passes through the point $R(2, 9)$ is

$$y - 9 = 8(x - 2)$$

which simplifies to

$$y = 8x - 7.$$

A point $P(x, y)$ is on L if and only if it satisfies the equation $y = 8x - 7$.

EXAMPLE 2. Find an equation for the line L passing through the points $R(16, 22)$ and $S(17, 21)$.

Solution. The slope of L is

$$m = \frac{22 - 21}{16 - 17} = -1.$$

Thus an equation for L is

$$y - 22 = -1(x - 16)$$

which simplifies to

$$x + y = 38.$$

A point $P(x, y)$ is on L if and only if it satisfies the equation $x + y = 38$.

Exercises

1. Find an equation for the line with slope $\frac{1}{2}$ which passes through $R(1, 4)$.
2. Find an equation for the line with slope $\frac{3}{4}$ which passes through $R(2, 3)$.
3. Find an equation for the line passing through $R(1, 2)$ and $S(4, 7)$.
4. Find an equation for the line pasing through $R(-1, -2)$ and $S(3, 4)$.

Appendix 2

The mathematics of finance

1. Simple and compound interest

In the world of finance, *interest* is simply money which is charged for the use of borrowed money. It is an amount which is stated in terms of some monetary unit (dollars, cents, pounds, francs, rubles). Thus if Mr. Debtor borrows $100 from Mr. Creditor with the understanding that $110 is to be repaid one month later, then the interest charged Mr. Debtor is $10.

Interest, money, should not be confused with interest rate, which is a number (no monetary units are involved). *Interest rate* is the ratio of the interest charged during the interest period to the amount of money owed at the beginning of the interest period. It is often stated as a percentage rather than in decimal form. The length of the interest period must be stated or understood. Thus Mr. Debtor's interest rate is

$$\frac{10}{100} = 0.10 \qquad \text{or} \qquad 10\% \text{ per month.}$$

Often, in colloquial language, the term interest rate is abbreviated to interest.

When the time came for Mr. Debtor to repay Mr. Creditor, he applied for an extension of the borrowing arrangement. This can be done in two ways. Mr. Debtor suggested that, in extending the arrangement for another month, interest be charged only on the amount originally borrowed ($100) so that the total interest to be paid is $2(\$10) = \$20$ and the amount to be paid at the end of the second month is

$$\$100 + 2(\$10) = \$120.$$

202

Mr. Debtor is suggesting a simple interest arrangement. If the interest due at the end of one interest period is c ($10 for Mr. Debtor) and the number of interest periods is n ($n = 2$ for Mr. Debtor), then

$$I = cn$$

is called *simple interest*. If P is the amount loaned (called the *principal*), then the *interest rate* r per interest period is

$$r = \frac{c}{p}$$

that is, the ratio of the interest due after one interest period to the amount originally loaned. Thus $c = rP$. Substitution of rP for c in $I = cn$ yields

$$I = Prn$$

as the simple interest due at the end of n interest periods if the principal P dollars is invested at the rate r per interest period. The sum A to be repaid at the end of n interest periods, called the *amount,* is principal plus interest.

$$A = P + Prn = P(1 + rn)$$

EXAMPLE 1.   Mr. Smith borrowed $1000 from Mr. Jones for three years at simple interest of 8% per annum. Find the interest and the amount.

Solution.   The rate 8% per annum tells us that the length of an interest period is one year. Thus $P = \$1000$, $r = 0.08$, $n = 3$, and it follows that

$$I = (\$1000)(0.08)(3) = \$240$$

and

$$A = \$1000 + \$240 = \$1240.$$

EXAMPLE 2.   Mr. Jones borrowed $2000 from Mr. Bank for 18 months at simple interest of 10% per annum. Find the interest and the amount.

Solution.   In this situation, $P = \$2000$, $r = 0.10$, and $n = \frac{3}{2}$.

$$I = (\$2000)(0.10)(\tfrac{3}{2}) = \$300$$
$$A = \$2000 + \$300 = \$2300$$

EXAMPLE 3.   A mortgage of $4000 is to be repaid in quarterly installments of $1000 plus simple interest at the rate of 8% per annum on the principal outstanding during each payment period. Find the total interest and the total amount.

Solution.   Four payments are to be made. The interest due with the first payment is ($4000)(0.08)($\tfrac{1}{4}$), or $80; the interest due with the second payment is ($3000)(0.08)($\tfrac{1}{4}$) = $60; the interest due with the third payment is ($2000)(0.08)($\tfrac{1}{4}$) = $40; the interest due with the fourth payment is ($1000)(0.08)($\tfrac{1}{4}$) = $20. Thus the total interest to be paid is $80 + $60 + $40 + $20 = $200 and the total amount paid is $4000 + $200 = $4200.

Returning to the negotiations between Mr. Debtor and Mr. Creditor, we see Mr. Creditor suggesting another basis for extending the borrowing arrangement. "At the end of the month," he points out, "you will owe me $110 consisting of the $100 principal and $10 interest. Thus in extending the loan another month it is reasonable to charge 10% interest on the $110 total owed instead of the $100 originally borrowed. Therefore the interest to be paid over the second month is ($110)(0.10) = $11; the total interest is $10 + $11 = $21, and the amount to be paid at the end of the second month is $121."

The arrangement favored by Mr. Creditor is one which entails adding the interest due to the principal owed, thereby forming an amount which serves as new principal with respect to which interest is determined for the next interest period. In general, if the interest owed is added to the principal at stated intervals of time, and thereafter itself earns interest as new principal, then the amount by which the original principal has been increased at the end of any time is called *compound interest*. The time between successive additions of interest to principal is called the *interest period*. The new principal at the end of an interest period, consisting of the principal of the previous period and the interest that has been added, is called the *compound amount*. In the borrowing arrangement proposed by Mr. Creditor to Mr. Debtor, the compound interest that would be owed at the end of two months is $21 and the compound amount due at the end of two months would be $121.

More generally, let us suppose that an initial amount of money A is invested at compound interest at a rate i per interest period. Then at the end of the first interest period, the compound amount is

$$A_1 = A + Ai = A(1 + i).$$

At the end of the second interest period, the new amount A_2 is A_1 plus the interest $A_1 i$ obtained from A_1.

$$A_2 = A_1 + A_1 i = A_1(1 + i) = A(1 + i)^2$$

At the end of the third interest period, the new amount A_3 is A_2 plus the interest $A_2 i$ obtained from A_2.

$$A_3 = A_2 + A_2 i = A_2(1 + i) = A(1 + i)^3$$

At the end of the fourth interest period, the new amount A_4 is A_3 plus the interest $A_3 i$ obtained from A_3.

$$A_4 = A_3 + A_3 i = A_3(1 + i) = A(1 + i)^4$$

More generally, at the end of the nth interest period, the new amount A_n is given by

$$A_n = A(1 + i)^n. \tag{1}$$

In financial transactions, an interest rate quoted as r *per annum compounded m times a year* is understood to mean that the year is divided into m interest periods of equal length and, for each of these interest periods, a rate of $i = r/m$ is used to determine the interest to be added to the compound amount at the end of a period. Thus 6% compounded two times a year (semiannually) means that the year is divided into two six-month periods with interest, determined at the rate of 3%, being added to the compound amount at the end of each six-month period. 12% per annum compounded twelve times a year means that the year is divided into twelve monthly periods with interest, determined at the rate of 1%, being added to the compound amount at the end of each monthly period.

If the interest rate is quoted as r per annum compounded m times a year, then $i = r/m$, and from equation (1) we obtain

$$A_n = A\left(1 + \frac{r}{m}\right)^n$$

as the compound amount after n interest periods. It is convenient to express n in terms of years. The number of interest periods in one year is m. Thus the number of periods in x years is mx. Therefore

$$C = A\left(1 + \frac{r}{m}\right)^{mx} \tag{2}$$

expresses the compound amount after x years if amount A is initially invested at a rate of r per annum compounded m times a year. x can take on integer values $(x = 0, 1, 2, \ldots)$ and fractional values which are consistent with m, that is, which divide evenly into m $(x = 1/m, 2/m, 3/m, \ldots)$. This result, equation (2), relates five quantities, namely, C, A, r, m, and x. If any four are known, the fifth can be determined.

EXAMPLE 4. If \$500 is invested at 8% per annum compounded quarterly (four times a year), what will it grow to in 4 years?

Solution. Since $A = \$500$, $r = 0.08$, $m = 4$, and $x = 4$, we have

$$C = 500(1.02)^{16}.$$

$(1.02)^{16} = 1.37279$; thus $C = (\$500)(1.37279) = \$686.40$.

If $C = A\left(1 + \dfrac{r}{m}\right)^{mx}$ is solved for A, we obtain

$$A = \frac{C}{\left(1 + \dfrac{r}{m}\right)^{mx}} = C\left(1 + \frac{r}{m}\right)^{-mx}$$

which tells us the amount that must initially be invested if amount C is to be attained mx years in the future, where the interest rate is r per annum compounded m times a year. In this setting A is called the *present value* of C. It expresses the present worth of an amount C to be realized in the future under the given growth conditions.

EXAMPLE 5. An investment is expected to realize $5000, 5 years from now. If the going interest rate is 9% per annum compounded 3 times a year, how much is the investment worth now?

Solution. Since $C = \$5000$, $r = 0.09$, $m = 3$, and $x = 5$, the present value of the investment is

$$A = (\$5000)(1.03)^{-15}.$$

$(1.03)^{-15} = 0.64186$, so that

$$A = (\$5000)(0.64186) = \$3209.30.$$

An initial investment of $3209.30 under the given interest conditions will yield $5000 five years from now.

2. Nominal and effective interest rates

Mr. Jones, an investor, is considering two investment possibilities. Option A guarantees an interest rate of 14% per annum compounded 14 times a year while option B guarantees an interest rate of 15% per annum compounded 3 times a year. Such rates are called nominal rates. More generally, the interest rate r per annum compounded m times a year which underlies compound interest arrangements is called the *nominal rate*. Mr. Jones' question is, Which of the two nominal rates described in options A and B yields the larger return?

One way of answering such a question is to state both nominal rates in terms of an interest rate based on a time period used as a standard and compare. The year is the time period usually used as a standard. For a given nominal rate r compounded m times a year, the corresponding *effective rate v* is that rate which, if compounded annually, would yield the same interest. The effective rate corresponding to the nominal rate 14% per annum compounded 14 times a year is, therefore, that rate which if compounded annually would yield the same interest. Mr. Jones' question can be answered by determining the effective rates corresponding to the nominal rates described in options A and B and comparing. We now turn to the problem of determining the effective rate v corresponding to a nominal rate r per annum compounded m times a year.

If interest is compounded once a year, the nominal rate and its corresponding effective rate are the same. If we equate the amounts at the end of one year under both arrangements, we obtain

$$A(1 + v) = A\left(1 + \frac{r}{m}\right)^m \tag{3}$$

where A is the initial amount invested. Dividing both members of (3) by A yields

$$1 + v = \left(1 + \frac{r}{m}\right)^m$$

or

$$v = \left(1 + \frac{r}{m}\right)^m - 1.$$

Thus the effective rate corresponding to a nominal rate of 14% per annum compounded 14 times a year is

$$v = \left(1 + \frac{0.14}{14}\right)^{14} - 1 = (1.01)^{14} - 1.$$

$(1.01)^{14} = 1.14947$; hence

$$v = 1.14947 - 1 = 0.14947 = 14.9\%.$$

This means that 14.9% compounded annually yields the same interest as 14% per annum compounded 14 times a year.

The effective rate corresponding to a nominal rate of 15% per annum compounded 3 times a year is

$$v = \left(1 + \frac{0.15}{3}\right)^{3} - 1 = (1.05)^{3} - 1$$
$$= 1.15762 - 1$$
$$= 0.15762$$
$$= 15.7\%.$$

Thus 15.7% compounded annually yields the same interest as 15% per annum compounded 3 times a year.

In terms of effective interest, then, option A versus option B can be stated as 14.9% compounded annually versus 15.7% compounded annually.

Appendix 3

A review of algebra

1. Exponents, roots, and radicals

If n is a positive integer, then x^n stands for the product $x \cdot x \cdot \cdots \cdot x$ with n factors. n, the number of factors in the product $x^n = x \cdot x \cdot \cdots \cdot x$, is called the *exponent* of x^n, and x^n itself is called the *nth power of x*. Thus, for example, $x^1 = x$, $x^2 = x \cdot x$, $x^3 = x \cdot x \cdot x$, $x^4 = x \cdot x \cdot x \cdot x$. In particular, x^2 is called the *square of* x, and x^3 is called the *cube of x*.

If there is a number x such $x^n = c$, where n is a positive integer greater than 1, then x is called an *nth root of c*. In particular, if $x^2 = c$, then x is called a *square root of c*. If $x^3 = c$, then x is called a *cube root of c*. Thus, for example, 4 and -4 are square roots of 16 since $4^2 = 16$ and $(-4)^2 = 16$; -2 is a cube root of -8 since $(-2)^3 = -8$; -2 and 2 are fourth roots of 16 since $(-2)^4 = 16$ and $2^4 = 16$. -4 does not have a square root within the real number system since there is no real number x such that $x^2 = -4$.

As the above examples make clear, a number c may have more than one *nth* root. One of these, called the *principal nth root*, is singled out by means of the following definition.

> If c is positive, its principal *nth* root is its positive *nth* root. If c is negative and n is odd (i.e., $n = 1, 3, 5, \ldots$), the principal *nth* root of c is the negative *nth* root of c.

The *radical sign* $\sqrt[n]{c}$ is used to denote the principal *nth* root of c; the number n written above the radical sign is called the *index* of the root. By definition, then,

$\sqrt[n]{c}$ has the property that

$$(\sqrt[n]{c})^n = c.$$

Thus, for example, the principal square root of 16, denoted by $\sqrt[2]{16}$ (or $\sqrt{16}$ with the index 2 omitted), is 4 since 4 is positive and $4^2 = 16$. If we wish to indicate the root -4 we write $-\sqrt[2]{16}$ (or $-\sqrt{16}$). It is common practice to omit the index 2 on the square root radical sign $\sqrt[2]{c}$ and simply write $\sqrt{c}$.

As further illustrations, let us observe that the principal cube root of -8, denoted by $\sqrt[3]{-8}$, is -2 since -2 is negative, n is odd, and $(-2)^3 = -8$. The principal fourth root of 16, denoted by $\sqrt[4]{16}$, is 2 since 2 is positive and $2^4 = 16$.

Exercises

Determine the value of each of the following.

1. $\sqrt{9}$ 2. $\sqrt[3]{-27}$ 3. $\sqrt[3]{-64}$ 4. $\sqrt{81}$

5. $\sqrt[3]{125}$ 6. $\sqrt[5]{32}$ 7. $\sqrt[4]{81}$ 8. $\sqrt[3]{-125}$

9. $(\sqrt{2})^2$ 10. $(\sqrt[3]{-10})^3$ 11. $(\sqrt[5]{c})^5$ 12. $(\sqrt[n]{a})^n$

The exponent concept is extended to negative integers, zero, and fractions in the following way. If $-n$ is a negative integer, then

$$x^{-n} = \frac{1}{x^n}, \quad \text{where } x \neq 0.$$

Thus $x^{-4} = \dfrac{1}{x^4}$, $\dfrac{1}{x^{-3}} = x^{-(-3)} = x^3$, and $3^{-2} = \dfrac{1}{3^2} = \dfrac{1}{9}$.

If $x \neq 0$, then x^0 is defined to be 1. Thus, $5^0 = 1$, $(2a)^0 = 1$ provided that $a \neq 0$. If p and q are integers, with q positive, then $x^{1/q}$ is defined by

$$x^{1/q} = \sqrt[q]{x}.$$

$x^{p/q}$ is defined by

$$x^{p/q} = (\sqrt[q]{x})^p.$$

Thus we have the following illustrations.

$x^{1/2} = \sqrt{x}$ $x^{1/3} = \sqrt[3]{x}$ $x^{1/4} = \sqrt[4]{x}$

$64^{1/2} = \sqrt{64} = 8$ $2^{1/3} = \sqrt[3]{2}$ $16^{3/2} = (\sqrt[2]{16})^3 = 4^3 = 64$

$27^{-1/3} = \dfrac{1}{27^{1/3}} = \dfrac{1}{\sqrt[3]{27}}$ $(-27)^{2/3} = (\sqrt[3]{-27})^2$ $x^{5/8} = (\sqrt[8]{x})^5$

$= (-3)^2 = 9$

$= \dfrac{1}{3}$

The following are basic properties of exponents.

1. $x^r \cdot x^s = x^{r+s}$

$$x^2 \cdot x^4 = x^{2+4} = x^6 \qquad x^{1/2} \cdot x^{3/2} = x^{(1/2)+(3/2)} = x^{4/2} = x^2$$
$$x^{2/3} \cdot x^{-1/3} = x^{(2/3)-(1/3)} = x^{1/3}$$

2. $(x^r)^s = x^{rs}$

$$(x^2)^4 = x^{2(4)} = x^8 \qquad (x^{1/3})^6 = x^{(1/3)6} = x^2$$

3. $(xy)^r = x^r \cdot y^r$

$$(xy)^4 = x^4 \cdot y^4 \qquad (xy)^{-1/2} = x^{-1/2} \cdot y^{-1/2}$$

4. If $y \neq 0$, $\left(\dfrac{x}{y}\right)^r = \dfrac{x^r}{y^r}$.

$$\left(\frac{x}{y}\right)^3 = \frac{x^3}{y^3} \qquad \left(\frac{x}{y}\right)^{-2} = \frac{x^{-2}}{y^{-2}} = \frac{\dfrac{1}{x^2}}{\dfrac{1}{y^2}} = \frac{y^2}{x^2}$$

5. If $x \neq 0$, $\dfrac{x^r}{x^s} = x^{r-s}$.

$$\frac{x^2}{x^4} = x^{2-4} = x^{-2} \qquad \frac{x^{1/2}}{x^{1/3}} = x^{(1/2)-(1/3)} = x^{1/6}$$

Exercises

Determine each of the following.

13. $64^{1/3}$ 14. $64^{-1/3}$ 15. $(-64)^{1/3}$
16. $(-64)^{-1/3}$ 17. $125^{2/3}$ 18. $125^{-2/3}$
19. $(-125)^{2/3}$ 20. $(-125)^{-2/3}$ 21. $16^{1/4}$
22. $16^{3/4}$ 23. $27^{-2/3}$ 24. $49^{1/2}$
25. $8^{4/3}$ 26. $(-8)^{-4/3}$ 27. $81^{-3/4}$

2. The cancellation principle and factoring

A fundamental principle of algebra is that a nonzero common factor may be introduced into or canceled from both the numerator and denominator of a fraction; that is,

$$\frac{a}{b} = \frac{ax}{bx}, \quad \text{where } x \neq 0.$$

When this principle is used to simplify an algebraic expression by canceling the common factor, it is often referred to as the *cancellation principle*. Its application presupposes that the numerator and denominator of the expression to be simplified have been expressed as products of factors, and this is where the operation called *factoring* enters the picture. To factor an algebraic expression means to express it as a product of other algebraic expressions.

EXAMPLE 1. Simplify $\dfrac{2t^2 - 8}{t - 2}$.

Solution. This problem arises in Exercise 10 of Section 1 and Examples 1 and 2 of Section 8. Because of the situations in this book in which this and certain other algebraic simplification problems like it arise, we are guaranteed that the denominator ($t - 2$ in this case) is one of the factors of the numerator ($2t^2 - 8$ in this case). This feature enormously simplifies our task. For the problem at hand our task is reduced to finding another factor of $2t^2 - 8$, where it is known that $t - 2$ is one such factor.

First, let us observe that since 2 is common to both terms of $2t^2 - 8$, it is a factor. We obtain

$$2t^2 - 8 = 2(t^2 - 4).$$

Since $t - 2$ is a factor of $2t^2 - 8$, it must also be a factor of $t^2 - 4$. What factor has the property that it times $t - 2$ yields $t^2 - 4$? That is our problem.

By inspection, $t + 2$ is suggested. This can be checked by multiplying $t - 2$ and $t + 2$ and observing whether or not $t^2 - 4$ is obtained. Since it is, we thus have

$$\frac{2t^2 - 8}{t - 2} = \frac{2(t^2 - 4)}{t - 2} = \frac{2(t - 2)(t + 2)}{(t - 2)}.$$

Since for $t \neq 2$, $t - 2$ is a common nonzero factor, it may be canceled from the numerator and denominator of $\dfrac{2(t - 2)(t + 2)}{(t - 2)}$ in accordance with the cancellation principle.

$$\frac{2t^2 - 8}{t - 2} = \frac{2(t - 2)(t + 2)}{(t - 2)} = 2(t + 2), \quad \text{where } t \neq 2$$

If inspection does not yield the other required factor, $t + 2$, then it can be obtained by dividing $t - 2$ into $2t^2 - 8$ or $t^2 - 4$. This approach is discussed in Example 3.

EXAMPLE 2. Simplify $\dfrac{600x - 5x^2 - 17,500}{x - 50}$.

Solution. This problem arises in a discussion of "A Situation in Economics" in Section 8. Let us first observe that -5 is common to all of the terms of $600x - 5x^2 - 17,500$.

$$600x - 5x^2 - 17,500 = -5(x^2 - 120x + 3500)$$

Since $x - 50$ is a factor of $600x - 5x^2 - 17{,}500$, it is also a factor of $x^2 - 120x + 3500$. By dividing 50 into 3500 we get the numerical term 70. Thus $x - 70$ or $x + 70$ are suggested possibilities for the desired other factor. Since the term 3500 is positive, $x - 70$ is the preferable candidate because the -50 of $x - 50$ multiplied by -70 of $x - 70$ yields 3500. It is a simple matter to verify that $(x - 50)(x - 70) = x^2 - 120x + 3500$. We thus obtain

$$\frac{600x - 5x^2 - 17{,}500}{x - 50} = \frac{-5(x^2 - 120x + 3500)}{x - 50}$$

$$= \frac{-5(x - 50)(x - 70)}{(x - 50)}$$

$$= -5(x - 70), \quad \text{where } x \neq 50.$$

If efforts of this sort to obtain the other required factor, $x - 70$, are unsuccessful, then the alternative is to divide $x - 50$ into $600x - 5x^2 - 17{,}500$ or $x^2 - 120x + 3500$. This approach is discussed in Example 4.

Division of polynomials

EXAMPLE 3.   Simplify $\dfrac{2t^2 - 8}{t - 2}$ by division.

Solution.   Since $2t^2 - 8 = 2(t^2 - 4)$, our problem is reduced to dividing $t^2 - 4$ by $t - 2$. To begin, the dividend $t^2 - 4$ is arranged in descending powers of t. Since no first-power t term is present in $t^2 - 4$, we insert a t term with a zero coefficient to systematize the division process. We thus start with the following.

$$t - 2 \overline{)\, t^2 + 0(t) - 4}$$

The first question to be answered is, What must t in the divisor $t - 2$ be multiplied by to yield the first term t^2 in the dividend? The answer, of course, is t. Multiplying both terms of the divisor $t - 2$ by t yields the following.

$$\begin{array}{r} t \phantom{{}^2 + 0(t) - 4} \\ t - 2 \overline{)\, t^2 + 0(t) - 4} \\ t^2 - 2t \end{array}$$

We next subtract.

$$\begin{array}{r} t \phantom{{}^2 + 0(t) - 4} \\ t - 2 \overline{)\, t^2 + 0(t) - 4} \\ t^2 - 2t \\ \hline 2t \end{array}$$

Bring down the next term, -4.

$$
\begin{array}{r}
t \phantom{{}+0(t)-4} \\
t - 2 \overline{\smash{)} t^2 + 0(t) - 4} \\
\underline{t^2 - 2t} \phantom{{}-4} \\
2t - 4
\end{array}
$$

We now repeat the process. What must t in the divisor $t - 2$ be multiplied by to yield $2t$? The answer, of course, is 2. Multiply both terms of the divisor $t - 2$ by 2.

$$
\begin{array}{r}
t + 2 \phantom{{}(t)-4} \\
t - 2 \overline{\smash{)} t^2 + 0(t) - 4} \\
\underline{t^2 - 2t} \phantom{{}-4} \\
2t - 4 \\
2t - 4
\end{array}
$$

Subtracting yields a remainder of zero.

$$
\begin{array}{r}
t + 2 \phantom{{}(t)-4} \\
t - 2 \overline{\smash{)} t^2 + 0(t) - 4} \\
\underline{t^2 - 2t} \phantom{{}-4} \\
2t - 4 \\
\underline{2t - 4} \\
0
\end{array}
$$

We thus have

$$
\frac{t^2 - 4}{t - 2} = t + 2, \quad \text{where } t \neq 2.
$$

The division can be checked by multiplication. Multiply $t - 2$ and $t + 2$.

$$
\begin{array}{r}
t - 2 \\
t + 2 \\
\hline
t^2 - 2t \\
+ 2t - 4 \\
\hline
t^2 \phantom{{}+2t} - 4
\end{array}
$$

Therefore

$$
\frac{2t^2 - 8}{t - 2} = \frac{2(t^2 - 4)}{t - 2} = 2(t + 2), \quad \text{where } t \neq 2.
$$

EXAMPLE 4. Simplify $\dfrac{600x - 5x^2 - 17{,}500}{x - 50}$ by division.

Solution. Since $\dfrac{600x - 5x^2 - 17{,}500}{x - 50} = \dfrac{-5(x^2 - 120x + 3500)}{x - 50}$ our problem is reduced to dividing $x^2 - 120x + 35$ by $x - 50$. The dividend $x^2 - 120x + 3500$ is arranged in descending powers of x, and the division is as shown below.

$$
\begin{array}{r}
x - 70 \\
x - 50 \overline{\smash{\big)}\; x^2 - 120x + 3500} \\
\underline{x^2 - 50x} \quad\quad\quad \text{subtract} \\
-70x + 3500 \\
\underline{-70x + 3500} \quad \text{subtract} \\
0
\end{array}
$$

We thus have

$$\frac{x^2 - 120x + 3500}{x - 50} = x - 70, \quad \text{where } x \neq 50.$$

Multiplying $x - 50$ and $x - 70$ as a check yields $x^2 - 120x + 3500$. Thus

$$\frac{600x - 5x^2 - 17{,}500}{x - 50} = \frac{-5(x^2 - 120x + 3500)}{x - 50} = -5(x - 70)$$

where $x \neq 50$.

EXAMPLE 5. Simplify $\dfrac{x^3 - a^3}{x - a}$ by division.

Solution. This problem arises in Example 8 of Section 9. First, the dividend $x^3 - a^3$ is arranged in descending powers of x. Since x^1 and x^2 terms are not present in $x^3 - a^3$, we insert them with zero coefficients to systematize the division process. Carrying out the division process yields the following development.

$$
\begin{array}{r}
x^2 + ax + a^2 \\
x - a \overline{\smash{\big)}\; x^3 + 0(x^2) + 0(x) - a^3} \\
\underline{x^3 - ax^2} \quad\quad\quad\quad\quad \text{subtract} \\
ax^2 + 0(x) \\
\underline{ax^2 - a^2x} \quad\quad\quad \text{subtract} \\
a^2x - a^3 \\
\underline{a^2x - a^3} \quad \text{subtract} \\
0
\end{array}
$$

Thus

$$\frac{x^3 - a^3}{x - a} = x^2 + ax + a^2, \quad \text{where } x \neq a.$$

Multiplying $x - a$ and $x^2 + ax + a^2$ as a check yields $x^3 - a^3$.

3. Logarithms

If b is a positive number, $b \neq 1$, and $b^y = x$, then the exponent y is called the *logarithm of x to the base b,* and we write

$$y = \log_b x.$$

Thus the logarithm of x to the base b is that exponent y to which b must be raised to obtain x.

$$\log_{10} 100 = 2 \quad \text{since} \quad 10^2 = 100$$
$$\log_2 8 = 3 \quad \text{since} \quad 2^3 = 8$$
$$\log_4 64 = 3 \quad \text{since} \quad 4^3 = 64$$
$$\log_{13} 1 = 0 \quad \text{since} \quad 13^0 = 1$$
$$\log_{12} 12 = 1 \quad \text{since} \quad 12^1 = 12$$

The use of logarithms as a tool for arithmetic computation is derived from the following three, well-known properties.

1. $\log_b MN = \log_b M + \log_b N.$

The logarithm of a product is the sum of the logarithms of the factors.

2. $\log_b \dfrac{M}{N} = \log_b M - \log_b N.$

The logarithm of a quotient is the logarithm of the numerator minus the logarithm of the denominator.

3. $\log_b M^r = r \cdot \log_b M.$

The logarithm of a factor raised to a power is the power times the logarithm of the factor.

Not as well known, perhaps, is the following property which is used in our discussion of derivatives of logarithmic functions in Section 15.

4. $\log_b N = \dfrac{1}{\log_N b}$

Proof. Let $y = \log_b N$ and $z = \log_N b$; then $b^y = N$ and $N^z = b$. Raising both members of $N^z = b$ to the y power yields

$$N^{zy} = b^y.$$

Since $b^y = N$ we have

$$N^{zy} = N.$$

Taking logarithms to the base N of each member of this equation yields

$$\log_N N^{zy} = \log_N N.$$

Since, by property 3, $\log_N N^{zy} = (zy) \cdot \log_N N$, and since $\log_N N = 1$, we have

$$zy \log_N N = \log_N N$$
$$zy = 1$$
$$y = \frac{1}{z}.$$

Since $y = \log_b N$ and $z = \log_N b$, substitution yields the desired result.

$$\log_b N = \frac{1}{\log_N b}$$

Answers to selected exercises

Chapter 1

2. $f(0) = f(-4) = f(3) = f(1) = 3$

4. $f(0) = f(1) = f(\frac{1}{2}) = 1$; $f(3) = f(4) = f(7) = f(24) = 2$

6. $f(0) = 0$, $f(\frac{1}{2}) = \frac{1}{4}$, $f(1) = 1$, $f(2) = 4$, $f(3) = 9$,
   $f(-\frac{1}{2}) = f(-1) = f(-\frac{3}{2}) = f(-2) = 3$

8. No, since $f(2)$ is not defined while $g(2) = 4$.

10. $g(t) = 2(t + 2)$, $t \neq 2$

12. $g(x) = \frac{1}{2}(x + 4)$, $x \neq 2$

14. $g(x) = 0$, $h(x) = 0$

16. $g(x) = \frac{1}{5}(x + 25)$, $x \neq 10$

18. $h(x) = \begin{cases} 2 + x^2, & x \geq 3 \\ 2, & x < 3 \end{cases}$

20.
$$T(x) = \begin{cases} 0.02x, & 0 < x \leq 1000 \\ 20 + 0.03(x - 1000), & 1000 < x \leq 3000 \\ 80 + 0.04(x - 3000), & 3000 < x \leq 5000 \\ 160 + 0.05(x - 5000), & 5000 < x \leq 7000 \\ 260 + 0.06(x - 7000), & 7000 < x \leq 9000 \\ 380 + 0.07(x - 9000), & 9000 < x \leq 11{,}000 \\ 520 + 0.08(x - 11{,}000), & 11{,}000 < x \leq 13{,}000 \\ 680 + 0.09(x - 13{,}000), & 13{,}000 < x \leq 15{,}000 \\ 860 + 0.10(x - 15{,}000), & 15{,}000 < x \leq 17{,}000 \\ 1060 + 0.11(x - 17{,}000), & 17{,}000 < x \leq 19{,}000 \\ 1280 + 0.12(x - 19{,}000), & 19{,}000 < x \leq 21{,}000 \\ 1520 + 0.13(x - 21{,}000), & 21{,}000 < x \leq 23{,}000 \\ 1780 + 0.14(x - 23{,}000), & 23{,}000 < x \leq 25{,}000 \\ 2060 + 0.15(x - 25{,}000), & 25{,}000 < x \end{cases}$$

22. $C(x) = 0.10\left(2\pi r^2 + \dfrac{24}{r}\right)$ 40. 4   42. 3   44. -5 46. 11   48. 4

50. 1   52. does not exist   54. does not exist   56. does not exist   58. 14

60. 83   62. 130/3   64. 4290   66. 15/17   68. 3588   70. 66/116

72. 464   74. 100   76. 3   78. 0   80. 7

82. does not exist   84. yes; $\lim\limits_{x\to 2} f(x) = 4$, $f(2) = 4$

86. yes, $\lim\limits_{x\to 1} f(x) = 3$, $f(1) = 3$ 88. no, $\lim\limits_{x\to 3} f(x)$ does not exist

90. yes, $\lim\limits_{x\to 3} f(x) = \frac{18}{13}$, $f(3) = \frac{18}{13}$ 92. yes, $\lim\limits_{x\to 0} f(x) = 0$, $f(0) = 0$

94. Yes, $f(x)$ is continuous at each real number c since
$\lim\limits_{x\to c} f(x) = c^2 + 2c + 1$ and $f(c) = c^2 + 2c + 1$.

96. $\lim\limits_{x\to\infty} f(x) = 0$ 98. $\lim\limits_{x\to 1} f(x) = \infty$

100. $\lim\limits_{x\to -2} f(x) = \infty$ 102. $\lim\limits_{x\to 0} f(x) = \infty$

104. $\lim\limits_{x\to -\infty} f(x) = 3/4$ 106. $\lim\limits_{x\to -1} f(x)$ does not exist

108. $\lim\limits_{x\to\infty} f(x) = 3$ 110. $\lim\limits_{x\to\infty} f(x) = 3/2$

112. a. 6.18%   b. 7.25%   c. 8.33%   d. 9.42%

114. $a(x) = 10{,}000\, e^{0.09x}$; $10{,}942$; $11{,}972$; $13{,}100$; $14{,}333$

116. $b(x) = 15{,}000\, e^{-0.06x}$; $13{,}303.50$; $12{,}529.50$; $11{,}799$

Chapter 2

2. 40   4. 5   6. 660   8. -3 10. 30/4   12. -40 14. $-1/9$

16. $f'(x) = -2/x^3$

18. If $f(x) = x^2 + 3x + 5$ is the time-distance function of an object in motion, then the instantaneous velocity of the object at time $x = 5$ is 13. The tangent line to the graph of $f(x) = x^2 + 3x + 5$ at the point $P(5, f(5)) = P(5,45)$ has slope 13. If $f(x) = x^2 + 3x + 5$ is the revenue function for the production of a certain commodity, then 13 is the marginal revenue for an output of 5 units. If $f(x) = x^2 + 3x + 5$ is the cost function for the production of a certain commodity, then 13 is the marginal cost for an output of 5 units.

20. $f'(0)$ does not exist.   22. $10x^9$ 24. $12x^{11}$ 26. $-3x^{-4}$ 28. $\frac{1}{5}x^{-4/5}$

30. $\frac{3}{4}x^{-1/4}$ 32. $\frac{1}{4}x^{-3/4}$ 34. $-\frac{3}{4}x^{-7/4}$ 36. $-5x^{-6}$ 38. $-10x^{-11}$

40. $3x^2 + 1$ 42. $2x + 1$ 44. $9x^8 + 5x^4$ 46. $6x^5 + 2x + \frac{1}{2}x^{-1/2}$

48. $(x^2 + x + 4)(4x^3) + (x^4 + 1)(2x + 1)$ 50. $24x^5 + 6x$ 52. $42x^2 - 7$

54. $6x^{-1/2} - 3$ 56. $(3x^2 - 3x + 1)(12x^2 - 4) + (4x^3 - 4x)(6x - 3)$

58. $(4x^3 + 3x^2 - 2x)(20x^3 - 12x) + (5x^4 - 6x^2 - 19)(12x^2 + 6x - 2)$

60. $(3x^6 + 4x^5 + 1)(8x - 6x^{-3}) + (4x^2 + 3x^{-2})(18x^5 + 20x^4)$

62. $\dfrac{(3x + 1)(6x^2) - (2x^3 - 4)3}{(3x + 1)^2}$

64. $\dfrac{(-4x^6 + 10)(9x^2) - (3x^3 + 9)(-24x^5)}{(-4x^6 + 10)^2}$

66. $\dfrac{(x^4 + 2x^2 - 7)(9x^2 + 2) - (3x^3 + 2x + 1)(4x^3 + 4x)}{(x^4 + 2x^2 - 7)^2}$

68. $\dfrac{(6x^7 - 9x^4 + 18)(108x^8 - 35x^4 - 4) - (12x^9 - 7x^5 - 4x + 15)(42x^6 - 36x^3)}{(6x^7 - 9x^4 + 18)^2}$

70. $6(3x^3 + 2x^2 + 8)^5(9x^2 + 4x)$

72. $\dfrac{1}{3}(x^6 + 2x + 2)^{-2/3}(6x^5 + 2)$

74. $90x^5(3x^2 + 1)^4 + (3x^2 + 1)^5(12x^3)$

76. $\dfrac{3(x^2 - 7)(x^3 + 5)^2(3x^2) - (x^3 + 5)^3(2x)}{(x^2 - 7)^2}$

78. $4(4x^5 + 2x - 8)^3(20x^4 + 2)$

80. $\frac{1}{3}(4x^3 + 2x - 1)^{-2/3}(12x^2 + 2)$

82. $\frac{175}{2}x^{10}(7x^5 - 13)^{-1/2} + \sqrt{7x^5 - 13}(30x^5)$

84. $\dfrac{6x\sqrt{3x - 7} - \frac{9}{2}x^2(3x - 7)^{-1/2}}{3x - 7}$

86. $x = y - 7$ 88. $y = \frac{1}{2}x^{1/3}$ 90. $x = \dfrac{1}{y - 1}$

92. $x = b^y$ 94. $\dfrac{dx}{dy} = \dfrac{1}{3}$ 96. $\dfrac{dy}{dx} = \dfrac{1}{3(y + 1)^2}$

98. $\dfrac{dx}{dy} = \dfrac{1}{3x^2}$ 100. $h(t) = A(b^t)$ 102. $h(t) = 3(3^t);\ 729$

104. $(\log_e 3)3^x$ 106. $4(\log_e 2)2^x + 3$ 108. $3\,e^x - 15x^2$

110. $4x^2 e^x + 8x e^x$ 112. $4\,e^{4x-1}$ 114. $240\,e^{0.12x}$

116. $-1500\,e^{-0.3x}$ 118. $100ae^{-ax}$ 120. $\dfrac{3(x^2 + 1)e^x - 6xe^x}{(x^2 + 1)^2}$

122. $(3x^4 + 2x)(\log_e 2)2^x + 2^x(12x^3 + 2)$

124. $3x^2 e^{x^3+2}$ 126. $k(\log_e A)(\log_e b)b^x A^{(b^x)}$

128. $500(\frac{1}{4}x^{-1/2} - 0.10)e^{\frac{1}{2}\sqrt{x} - 0.01x}$ 130. $\dfrac{\log_5 e}{x}$

132. $4x^2 + (\ln x)(12x^2)$ 134. $\dfrac{(5x^3 + 4x + 7)}{x} + (\ln x)(15x^2 + 4)$

136. $\dfrac{e^x}{x} + e^x \ln x$ 138. $\dfrac{10x}{5x^2 - 19}$ 140. $\dfrac{5(\log_3 e)}{5x + 1}$

142. $\dfrac{2xe^{-x}}{x^2 + 1} - e^{-x} \ln(x^2 + 1)$ 144. $\dfrac{3(\ln x)^2}{x}$ 146. $\dfrac{-3y}{x}$

148. $\dfrac{-8xy - 4}{4x^2}$ 150. $\dfrac{-x}{y}$ 152. $\dfrac{-y - 2x}{x}$ 154. $\dfrac{-x}{y}$ 156. $\dfrac{2}{y}$

158. $c'(x) = \frac{1}{4}x^2 - 2x + 25;\ c'(12) = 37;\ \kappa = \frac{111}{275}$

160. $\eta = 1.8;\ R(x) = \dfrac{800x}{x + 10} - 4x;\ R'(x) = \dfrac{8000}{(x + 10)^2} - 4;\ R'(10) = 16$

164. $\eta = \dfrac{2}{21};\ R(x) = x(50 - 2x)^2;\ R'(x) = 12x^2 - 400x + 2500;\ R'(21) = -608$

166. 12   168. 6   170. 1/4   172. 1/50   174. 1/5000   176. -86

Answers to selected exercises

Chapter 3

2. $f(2) = 17$ is a maximum value.
4. $f(-3) = -11/2$ is a minimum value.
6. $f(0) = 0$ is a minimum value; $f(-1) = 1$ is a maximum value.
8. $f(-3) = 25/2$ is a maximum value and $f(2/3) = -328/27$ is a minimum value.
10. $f(-2) = 10/3$ is a maximum value and $f(3) = -35/2$ is a minimum value.
12. $f(0) = 1/\sqrt{2\pi}$ is a maximum value.
14. $x = 100$ 16. $x = 450$ 18. $x = 50$ 20. $x = 57$ 22. $x = 530$
24. height equals diameter   26. 7.5%   28. 5   30. 100 foot front, 200 foot side
32. P is $40/3$ miles from A.
34. a. $x = 100$ tons; \$57,000   b. $x = 90$ tons is the after-tax optimal output.
    c. \$10,800   d. \$45,600   e. \$46,200
    f. The market price rises from \$704 per ton to \$754 per ton.
36. a. $x = 40$ tons; \$28,867   b. $x = 35$ tons is the after-tax optimal output.
    c. \$10,325   d. \$17,783   e. \$18,542
    f. The market price rises from \$1020 per ton to \$1025 per ton.
38. a. \$600 per ton   b. $x = 50$ tons is the after-tax optimal output.   c. \$30,000
    d. \$12,000   e. \$27,000
    f. The market price rises from \$704 per ton to \$954 per ton.
40. 6.25 years

Chapter 4

2. $\bar{c}(x)$ decreases for $0 < x < 2$ and increases for $x > 2$. $\bar{c}(x)$ is concave upward for $x > 2$. There are no inflection points.
4. $P(x)$ increases for $0 < x < 450$ and decreases for $x > 450$. $P(x)$ is concave downward for $x > 0$. There are no inflection points.
6. $f(x)$ increases for $x < -3$, decreases for $-3 < x < \frac{2}{3}$, and increases for $x > \frac{2}{3}$. $f(x)$ is concave downward for $x < -\frac{7}{6}$ and concave upward for $x > -\frac{7}{6}$. $(-\frac{7}{6}, \frac{29}{216})$ is an inflection point.
8. $f(x)$ increases for $x < -2$, decreases for $-2 < x < 3$, and increases for $x > 3$. $f(x)$ is concave downward for $x < \frac{1}{2}$ and concave upward for $x > \frac{1}{2}$. $(\frac{1}{2}, -\frac{85}{12})$ is an inflection point.
10. $f(x)$ increases for $x < 1$ and decreases for $x > 1$. $f(x)$ is concave downward for $x < 2$ and concave upward for $x > 2$. $(2, 2/e^2)$ is an inflection point.
12. $\bar{c}(x)$ is continuous for $x > 0$; $(2, \frac{5}{2})$ is a minimum point; $\bar{c}(x)$ decreases for $0 < x < 2$ and increases for $x > 2$; $\bar{c}(x)$ is concave upward for $x > 0$. There are no inflection points. $\lim_{x \to 0} \bar{c}(x) = \infty$ and $\lim_{x \to \infty} \bar{c}(x) = \infty$. $\bar{c}(x)$ lies entirely above the x-axis.
14. $P(x)$ is continuous for $x > 0$; $(450, 202,490)$ is a maximum point; $P(x)$ increases for $0 < x < 450$ and decreases for $x > 450$; $P(x)$ is concave downward for $x > 0$. There are no inflection points. $\lim_{x \to 0} P(x) = -10$ and $\lim_{x \to \infty} P(x) = -\infty$.
16. $P(x)$ is continuous for $x > 0$; $(501, 250,901)$ is a maximum point; $P(x)$ increases for $0 < x < 501$ and decreases for $x > 501$; $P(x)$ is concave downward for $x > 0$. There are no inflection points. $\lim_{x \to 0} P(x) = -100$ and $\lim_{x \to \infty} P(x) = -\infty$.
18. $f(x)$ is continuous for all x; $(-3, \frac{25}{2})$ is a maximum point and $(\frac{2}{3}, -\frac{328}{27})$ is a minimum point; $f(x)$ increases for $x < -3$, decreases for $-3 < x < \frac{2}{3}$, and increases for $x > \frac{2}{3}$.

$f(x)$ is concave downward for $x < -\frac{7}{6}$ and concave upward for $x > -\frac{7}{6}$; $(-\frac{7}{6}, \frac{29}{216})$ is an inflection point. $\text{Limit}_{x \to \infty} f(x) = \infty$ and $\text{limit}_{x \to \infty} f(x) = -\infty$.

20. $f(x)$ is continuous for all x; $(-2, \frac{10}{3})$ is a maximum point and $(3, -\frac{35}{2})$ is a minimum point; $f(x)$ increases for $x < -2$, decreases for $-2 < x < 3$, and increases for $x > 3$. $f(x)$ is concave downward for $x < \frac{1}{2}$ and concave upward for $x > \frac{1}{2}$; $(\frac{1}{2}, -\frac{85}{12})$ is an inflection point. $\text{Limit}_{x \to \infty} f(x) = \infty$ and $\text{limit}_{x \to \infty} f(x) = -\infty$.

22. $f(x)$ is continuous for all x; $(1, 1/e)$ is a maximum point; $f(x)$ increases for $x < 1$ and decreases for $x > 1$. $f(x)$ is concave downward for $x < 2$ and concave upward for $x > 2$; $(2, 2/e^2)$ is an inflection point. $f(0) = 0$.

Chapter 5

2. $\frac{1}{11}x^{11} + C$ 4. $-\frac{1}{6}x^{-6} + C$ 6. $\frac{4}{7}x^{7/4} + C$ 8. $\frac{5}{2}x^{6/5} + C$

10. $-8x^{-1/2} + C$ 12. $\frac{1}{3}x^3 + x + C$ 14. $\frac{2}{5}x^5 + \frac{3}{2}x^2 + x + C$

16. $\frac{8}{3}x^{3/2} + \frac{3}{2}x^{-2} + C$ 18. $9x^{1/3} - 4e^x + C$ 20. $\frac{5}{7}x^7 - \ln x + C$

22. $-(1/x) + \ln x + C$ 24. $\frac{1}{5}(x^2 + 1)^5 + C$ 26. $\frac{1}{3}(1 + x^2)^{3/2} + C$

28. $10(x^2 + 1)^{1/2} + C$ 30. $-4000 e^{-0.5t} + C$ 32. $\ln(x^2 - 1) + C$

34. $\frac{1}{6}(10 + x^4)^{3/2} + C$ 36. $-20,000 e^{0.1(10-t)} + C$ 38. $-(A/r)e^{r(x-t)} + C$

40. $\frac{1}{2}\ln(x^2 + 1) + C$ 42. $-\frac{1}{2}e^{-x^2} + C$ 44. $-(5 - x^2)^{3/2} + C$ 46. $\frac{1}{3}e^{x^3} + C$

48. $(x/3)e^{3x} - \frac{1}{9}e^{3x} + C$ 50. $-12,000 e^{-0.05t} + 400te^{-0.05t} + C$

52. $-1300 e^{0.1(5-t)} - 30te^{0.1(5-t)} + C$ 54. $x \ln x - x + C$

56. $\frac{1}{2}x^2 \ln x - \frac{1}{4}x^2 + C$ 58. $-x^2e^{-x} - 2xe^{-x} - 2e^{-x} + C$

60. $y = \frac{2}{3}x^3 + \frac{3}{2}x^2 + x + \frac{31}{6}$ 62. $c(x) = \sqrt{100x + 2000} + 105.28$

64. $R(x) = 100x - 5x^2$ 66. $y = t^2 + 2t$ 68. $I(t) = 1000 e^t - 50t^2 + 18,000$; $I(2) = \$25,189.10$

70. $\sum_{i=1}^{i=4} (1 - \sqrt{i})$ 72. $\sum_{i=1}^{i=4} \left(i + 3 - \frac{i}{i+1}\right)$ 74. $\sum_{i=1}^{i=n-1} x_i x_{i+1}$

76. No; 4.99 is not the least upper bound since it is not an upper bound. 4.991, for example, is in the set and is larger than 4.99. 1.99 is not the greatest lower bound since 2 is also a lower bound and 2 is greater than 1.99.

78. 1   80. a. $150 \le A(S) \le 582$ b. $160 \le A(S) \le 568$ c. $240 \le A(S) \le 456$

82. a. $17 \le W \le 34$ b. $22 \le W \le 31$

84. 28/3   86. 10/3   88. 10   90. 3/2   92. 44,000/3   94. 93

96. 231,855/4   98. $1000(1 - e^{-4})$ 100. e 102. $14,000e^{-1} - 6000$

104. $1300 e^{0.5} - 1450$ 108. 128/3   110. 72   112. 211/4   114. 50/3

116. $\frac{14}{3} - \ln 4$ 118. 5/12   120. $e^3 + e^{-3} - 2$ 122. a. 36,000   b. 11,164/3

c. 10,444/3   124. a. 4000/3   b. 972   126. \$12,642   128. \$17,726

130. a. \$10,057   b. \$9294   132. \$34,366   134. \$88,181   136. \$693

138. 1720/3 ft-pds   140. 16,596 ft-pds   142. 1/3   144. 1   146. 1   148. 1

150. does not exist   152. \$10,000

Chapter 6

2. $f(2, 3, 1) = 21$; $f(-1, 2, 4) = 12$; $f(3, 2, -1) = 8$

4. a. $f(x, 1) = 3x^2 + x$; $f_x(3, 1) = 19$ b. $f(x, 3) = 3x^2 + 9x$; $f_x(2, 3) = 21$

c. $f(2, y) = 2y^2 + 12$; $f_y(2, 4) = 16$ d. $f(-1, y) = -y^2 + 3$; $f_y(-1, 3) = -6$

6. $f_x(x,y) = 6x + y^2$; $f_y(x,y) = 2xy$

8. $f_x(x,y) = 12x^2y^2 + 6x + y^2$; $f_y(x,y) = 8x^3y + 2xy$; $f_x(1,2) = 58$; $f_y(2,-1) = -68$

10. $f_x(x,y,z) = ye^z$; $f_y(x,y,z) = xe^z$; $f_z(x,y,z) = (xy)e^z$

12. $f_x(x,y,z) = \dfrac{(3x + 2y + 4z)(2yz^3) - 6xyz^3}{(3x + 2y + 4z)^2}$

$f_y(x,y,z) = \dfrac{(3x + 2y + 4z)(2xz^3) - 4xyz^3}{(3x + 2y + 4z)^2}$

$f_z(x,y,z) = \dfrac{(3x + 2y + 4z)(6xyz^2) - 8xyz^3}{(3x + 2y + 4z)^2}$

14. $f_x(x,y) = \dfrac{x}{\sqrt{x^2 + y^2}}$; $f_y(x,y) = \dfrac{y}{\sqrt{x^2 + y^2}}$

16. $f_x(x,y,z) = \dfrac{(2xy + z)(2xz^2) - 2x^2yz^2}{(2xy + z)^2}$; $f_y(x,y,z) = \dfrac{-2x^3z^2}{(2xy + z)^2}$

18. $f_{xx}(x,y) = 18x$; $f_{yy}(x,y) = 6xy$; $f_{xy}(x,y) = f_{yx}(x,y) = 3y^2$; $f_{xx}(1,2) = 18$;
$f_{yy}(-1,3) = -18$; $f_{yx}(4,1) = 3$

20. $f_{xx}(x,y) = 12xy^2 - 2y^4$; $f_{yy}(x,y) = 4x^3 - 12y^2x^2$;
$f_{xy}(x,y) = f_{yx}(x,y) = 12x^2y - 8y^3x$; $f_{xx}(-1,2) = -80$; $f_{yy}(1,3) = -104$;
$f_{xy}(2,3) = -288$; $f_{yx}(4,1) = 160$

22. $f_{xx}(x,y,z) = 80x^3y^3 + 6xz^4$; $f_{yy}(x,y,z) = 24x^5y$; $f_{zz}(x,y,z) = 12x^3z^2$;
$f_{xy}(x,y,z) = f_{yx}(x,y,z) = 60x^4y^2$; $f_{xz}(x,y,z) = f_{zx}(x,y,z) = 12x^2z^3$;
$f_{yz}(x,y,z) = f_{zy}(x,y,z) = 0$

24. $f(0,0) = 0$ is a minimum value.

26. $f(-1,0) = 2$ is a maximum value; $(1,0)$ is a critical point which does not yield an extreme value.

28. There are no extreme values. $(0,0)$ is a critical point which does not yield an extreme value.

30. There are no extreme values. $(-1,-1)$ is a critical point which does not yield an extreme value.

32. $f(2,0) = -16$ is a minimum value; $f(-2,0) = 16$ is a maximum value. $(0, \sqrt{12})$ and $(0, -\sqrt{12})$ are critical points which do not yield extreme values.

34. $x = y = z = 2$

Index

A B C D E F G H   7 9 8 7 6 5